THE CORNISH MINER IN AUSTRALIA:
Cousin Jack Down Under

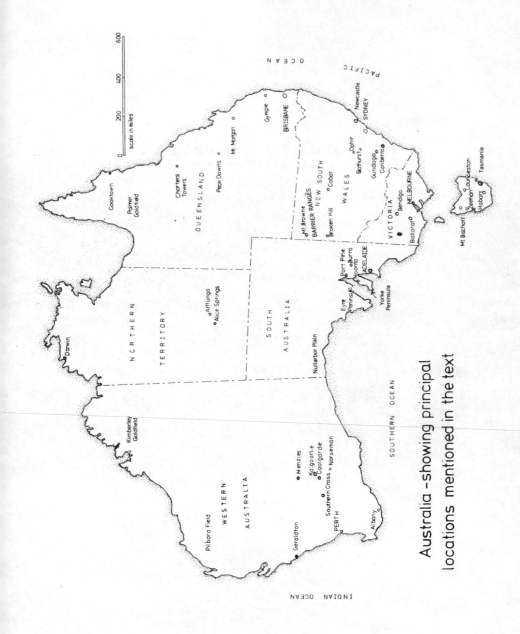

Australia –showing principal locations mentioned in the text

The Cornish Miner
in
Australia

(Cousin Jack Down Under)

Philip J Payton BSc PhD

(Bardic Name: Car Dyvresow)

DYLLANSOW TRURAN

First published in 1984
By Dyllansow Truran
Trewolsta, Trewirgie, Kernow, [Cornwall]

Printed in Great Britain
by Penryn Litho, Mabe, Penryn, Kernow
ISBN 0 907566 51 0 [Hardback]
0 907566 52 9 [Paperback]

By the same author:

Pictorial History of Australia's Little Cornwall

The Story of HMS Fisgard

Cornish Carols From Australia (forthcoming)

For Brigid Angharad

ACKNOWLEDGEMENTS

Without the advice, assistance and encouragement of a great many individuals this book never would have been written. I was inspired initially by the superb examples already set by A.L. Rowse, A.C. Todd and John Rowe, with their various volumes on the Cornish in North America, and their tremendous enthusiasm for my project helped considerably in those difficult, early days of commencing research. The great interest shown by Charles Thomas and Geoffrey Blainey was similarly encouraging. The pioneer work of Geoffrey Blainey, Oswald Pryor, D. Bradford Barton and Ian Auhl was of vital importance in suggesting directions in which my own research might develop, as were the more recent efforts by Mandie Robinson and (in the case of New Zealand, which, alas, I have been barely able to consider) Audrey Paterson. I am also indebted, as the bibliography well shows, to a great many other authors of Cornish and Australian titles, and also to the many who over the years have deposited theses, manuscripts, letters, diaries, photographs and other historical material in museums and archives for the benefit of research workers such as myself.

A number of people were kind enough to allow me access to their personal papers and collections, and here I would especially like to thank John Cowling, E.M. Cunnack, V.M. Easterbrook and Henry Giles for their permission to make use of private materials that they hold. During my time in Australia I was helped by numerous individuals, but those deserving particular mention are my former tutor, John Playford, John Tregenza, who allowed me to inspect the significant material that he had collected, May Cocks, Roslyn Paterson and Ian Auhl. I should also like to thank the various museums, libraries, universities and archives which afforded me help and allowed me to use their material - in particular I should like to thank the long suffering but ever tolerant staff of the Archives Section of the State Library of South Australia who had to endure my awkward requests and difficult questions for so many months. The photographs in this book are reproduced courtesy of the State Library of South Australia, the La Trobe Library, the Charles Rasp Memorial Library, and the Royal Institution of Cornwall, as detailed in the captions. The line drawings from Sherer's *Gold-Finder of Australia* are reproduced by permission of Penguin Books (Australia) from their recent facsimile edition of the volume.

I am, of course, greatly indebted to the University of Adelaide, which provided the opportunity for me to write my PhD thesis, and I am similarly indebted to my publisher - Len Truran - for enabling the fruits of my Australian sojourn to at last see the light of day. And finally I must say thank you to Jane, my wife, who not only tolerated (and indeed assisted) three years full-time research but had also, more recently, to put up with the devotion of a at least some of my leave periods to the "re-write."

Philip Payton

FOREWORD

Dr Philip Payton's PhD thesis, "The Cornish in South Australia", has come to be regarded as something of a classic. The copies deposited in the State and University libraries in Adelaide are in constant demand from both amateur and professional historians, readers ranging from family history enthusiasts to university students and academics. The thesis has become *the* standard work on the Cornish in this continent, and I am delighted that it has now been re-worked into a published book which will be available widely in Australia and the United Kingdom.

During his time in Australia, Philip Payton became a familiar figure in Adelaide's Archives and Libraries. He also travelled widely in his quest for the Cornish, not only visiting Kapunda, Burra, Moonta and the other South Australian sites but also venturing much further afield — going underground in the South Mine at Broken Hill, exploring the goldfields of New South Wales, meeting Geoffrey Blainey in Melbourne to compare notes about the Cousin Jacks in Victoria. Philip was also active in the Cornish Association of South Australia and the Historical Society of South Australia, and — in addition to his teaching responsibilities at the University and the Institute of Technology — lectured extensively to local schools and tertiary institutions. He also wrote a book entitled *Pictorial History of Australia's Little Cornwall* which was published by Rigby Ltd in 1978 and won an Australia and New Zealand Bank literary prize.

Philip's work reflects his broad inter-disciplinary background in the social sciences. He displays sound historical judgement, but also exhibits a keen sense of geography and "place", his natural affinity with Cornwall almost matched by a strong and intimate feeling for Australia. And he is politically aware, his appreciation of the important role played by the Cornish in the rise of the United Labor Party and its associated trade unions fills a significant gap in South Australian Labor history.

In so many ways a pioneer study, Philip Payton's work should prove an impetus to further research into Cornwall's "Great Migration" and the impact of the Cornish overseas, much of which is left still to be chronicled. My own *Cornish in Australia* (a brief summary of several sources broken down for local secondary school consumption) draws to a considerable degree upon the inspiration and material provided in his PhD thesis.

Others, perhaps, will follow with volumes on the Cornish in New Zealand, Canada, South America and elsewhere, and I understand that Philip has completed already a study of *Cornish Carols From Australia*, with a further volume on Cornish involvement in the expansion of the South Australian agricultural frontier at present in the pipeline. Philip's enthusiasm for Australia is to be applauded — may its literary expression continue to tell us more about our Cousin Jack heritage.

Jim Faull,
Senior Vice-President,
Cornish Association of South Australia,
Senior Lecturer in Geographical Studies,
South Australian College of Advanced Education.
September 1983

CONTENTS

Acknowledgements	vi
Foreword	vii
List of Illustrations	xi
Introduction	1
Beginnings	7
Part 1: The Central Colony	11
Chapter 1 — The Great Migration	12
Chapter 2 — The Mineral Kingdom	32
Chapter 3 — Little Cornwall	52
Chapter 4 — The Radical Tradition	89
Part 2: The Diaspora	109
Chapter 5 — The Lure of Gold	110
Chapter 6 — The Barrier and the Hill	135
Chapter 7 — The Golden West	174
Chapter 8 — Far and Wide	189
And Endings	202
Appendix A	213
Appendix B	215
Appendix C	216
Select Bibliography	217
Newspapers & Periodicals	219
Books, Pamphlets & Articles	220
University of Adelaide Unpublished BA. Honours & other Theses	224
Miscellaneous Sources	224
Notes	225
Authors Postscript	229
Index	231

LIST OF ILLUSTRATIONS

	Page
Hughes' engine and engine-house, Moonta Mines	Front Cover
Map of Australia showing principal locations	ii
Map of South Australia	2
Map of principal Victorian Goldfields	3
Map of New South Wales Goldfields, ~~1850~~s	4
Map of Cornwall	5
"Emigration to South Australia"	53
"Free Emigration to Port Adelaide"	54
The Hooghly at Port Adelaide	55
Philip Santo, from Saltash	56
James Crabb Verco, from Callington	57
Kapunda Mine	58
A section of Kapunda Mine	58
A panoramic view of the Burra Burra Mine	59
Wheal Blinman - The Blinman Mine	60
Blinman Township	61
Elder's engine-house, Wallaroo Mines	62
An early mine on Yorke peninsula	63
Cornish miners at Hughes' Shaft, Moonta Mines	64
Unloading ore, Wallaroo Mines	65
Miners at a shaft plat, Wallaroo Mines	66
A double-decked man-skip, Wallaroo Mines	67
Methodist Chapel at Cross Roads, near Moonta	68
One of the first cottages built at Moonta Mines	149
Sir James Penn Boucaut	150
Strike leaders and their supporters at Moonta and Wallaroo	151
"Forest Creek", Victoria	152
Nicholls' "Diggers Executing Their Own Laws"	153
Nicholls' "Successful At Last"	154
"The Unlucky Digger That Never Returned"	155
"The Invalid Digger"	156
"Preaching At The Diggings"	157
The "Welcome Stranger" Nugget	158
Argent Street, Broken Hill	159
A view along "the line of the lode"	160
The Great Boulder Mine	161
The Great Boulder Mine	161
"Entirely Free Emmigration to Van Diemen's Land and New South Wales"	162
"To Miners...An Early Free Passage to New Zealand"	163
"Honest John" Verran, from Gwennap	164
Callington Mine, South Australia	Back Cover

INTRODUCTION

This book is the culmination of some three years research and writing in Australia, from late 1975 until the end of 1978, when I produced a PhD thesis entitled "The Cornish in South Australia: Their Influence and Experience From Immigration to Assimilation 1836-1936" at the University of Adelaide. The thesis was intended originally as an in-depth and at-length examination of the Cornish impact in one specific and relatively confined geographic area - a sort of microcosm of the Cornish emigrant experience across the world. But, while this was indeed accomplished, two important facts which I had only half-anticipated emerged strongly from the research: that the Cornish influence in South Australia was due principally to the immigration of copper miners, and that - perhaps most importantly - the Cornish mining impact in other parts of the continent was due in the first place to the dynamism of South Australia's Cousin Jacks who, for example, pioneered the growth of Broken Hill in the 1880s and were amongst the very first gold-seekers in Victoria in 1851 and on the West Australian fields forty years later.

To be sure, it was in the mining towns of the "Central Colony" (as South Australia was often known) that the Cornish identity was most deeply imbedded and the Cornish atmosphere most keenly felt. Milton Hand, an Australian local historian, described Moonta, Wallaroo, and Kadina as "... three towns which... provided the State with its unique Cornish heritage." And Ian Auhl - in assessing the extent of the Cornish impact - argued that "What happened in the old mining towns of South Australia, at Kapunda, Burra, Moonta, Wallaroo, and Kadina, is unique in Australia ......" Our own A.L. Rowse has added that the Cornishman's "... mark is strong upon Australia (particularly South Australia) ..."

But having said that, it must be admitted that the Cornish influence was felt - to a greater or lesser degree - across the continent, in metaliferous mining fields from Kalgoorlie in Western Australia to Cobar in New South Wales, and from Peak Downs in Queensland to Ballarat in Victoria and Mount Bischoff in Tasmania. This book, therefore, while devoting principal attention to the Central Colony (in PART I), ranges in content across Australasia (in PART II) - with areas as far-flung as New Zealand, New Caledonia, and even New Guinea receiving at least some attention.

Philip Payton,
Torpoint, Cornwall.
January 1983.

1

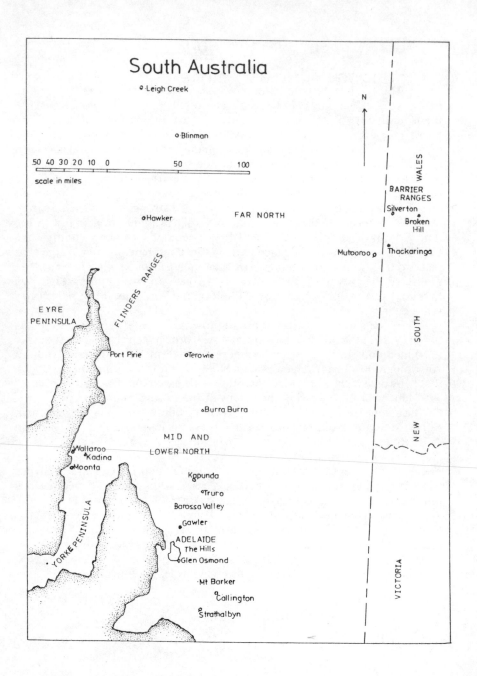

South Australia

o Leigh Creek

o Blinman

50 40 30 20 10 0   50   100

scale in miles

N

FAR NORTH

BARRIER
RANGES

Silverton

Broken
Hill

o Hawker

Mutooroo o | Thackaringa

FLINDERS RANGES

EYRE
PENINSULA

Port Pirie

o Terowie

o Burra Burra

MID AND
LOWER NORTH

Wallaroo
Kadina
Moonta

Kapunda

o Truro

Barossa Valley

o Gawler

ADELAIDE
The Hills
Glen Osmond

YORKE PENINSULA

Mt Barker

o
Callington

o
Strathalbyn

WALES

SOUTH

NEW

VICTORIA

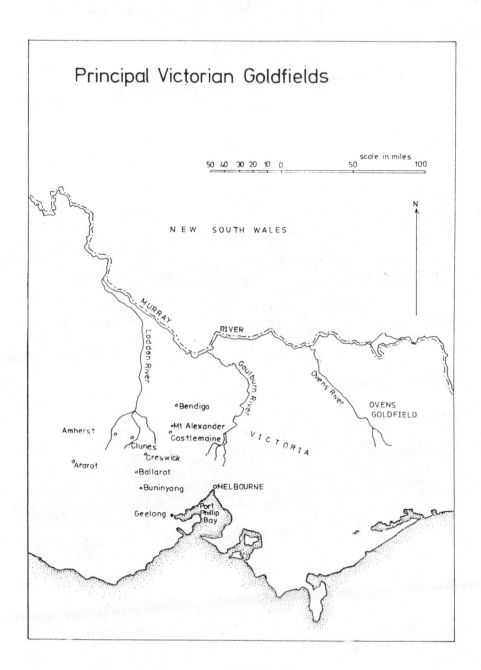

Principal Victorian Goldfields

scale in miles

50 40 30 20 10 0   50   100

N

NEW SOUTH WALES

MURRAY   RIVER

Loddon River

Goulburn River

Ovens River

OVENS
GOLDFIELD

Amherst

Bendigo

Mt Alexander
Castlemaine

VICTORIA

Clunes
Creswick

Ararat

Ballarat

Buninyong   MELBOURNE

Geelong   Port
Phillip
Bay

3

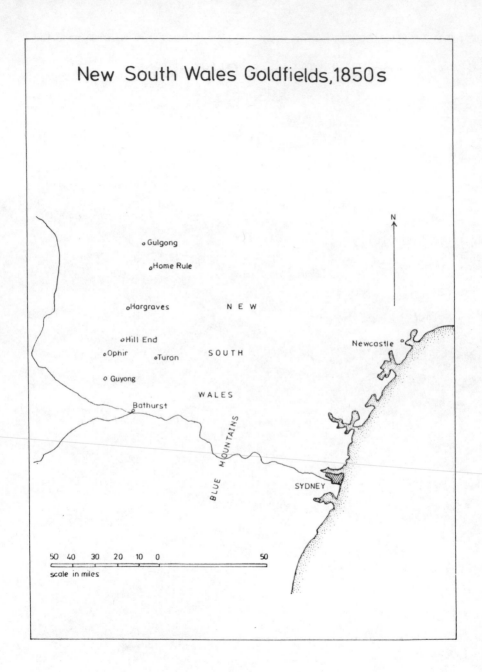

New South Wales Goldfields, 1850s

o Gulgong

o Home Rule

o Hargraves   N E W

o Hill End
o Ophir   o Turon   S O U T H

o Guyong

Newcastle o

W A L E S

Bathurst

BLUE MOUNTAINS

SYDNEY

50 40   30   20   10   0   50

scale in miles

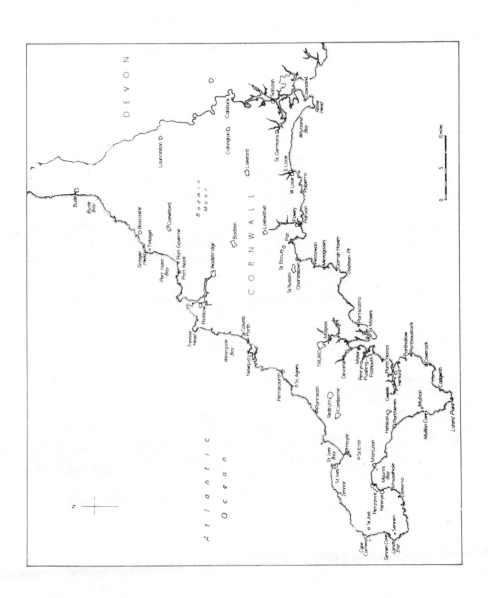

BEGINNINGS...

The distinctive background of Cornwall is too well-known to be related here in any detail, but nevertheless a knowledge of certain of its elements is essential if the influence and experience of the Cornish miners in Australia is to be understood. An appreciation of Cornwall's Celtic identity is crucial - Cornwall is the land of Trystan and Arthur, of Piran and Petroc, of the Padstow 'Obby 'Oss and St. Columb Hurling, of the Furry Dance and Cornish Wrestling. Even until modern times it had its own Tinners' Parliament and its own language, elements of both surviving to shape the character of present-day Cornwall, and in the last century the saying "into Cornwall, out of England" was well-known. Cornwall was geographically remote from the rest of Britain, a sea-girt land of granite and moor thrust into the Atlantic, and was considered a place apart by both the Cornish themselves and the "foreigners" who lived on the otherside of the River Tamar border. In several respects the Cornish had more in common with their cousins, the Bretons and Welsh, than with the people of neighbouring Devon.

But although remote, Cornwall was a birth-place and cradle of the Industrial Revolution. Steam power was first put to viable technological use in the Cornish mines, and for more than a century Cornwall was hailed as the mining centre of the world. Physically, mining dominated the environment of much of Cornwall - from Gunnislake to St. Just-in-Penwith - the mining districts clearly defined by the towering walls and lofty stacks of the Cornish engine-houses. Often the grandeur of these buildings was accentuated by their location - on the crest of a windswept hill or perched on the edge of a cliff with the sea pounding a hundred feet below. But just as the architecture of Cornish mining was heroic, so the economic fortunes of the industry were volatile - the 1860s witnessing the crumbling of Cornwall's copper empire, and tin experiencing near-disaster less than a decade later.

Farming in Cornwall, too, suffered periodic depressions (as in the 1840s and 1870s), and Cornish fishing experienced a gradual decline in the last century, culminating in crisis in the 1890s. Social conditions, not surprisingly, reflected economic conditions - combining with a bleak environment and the foreign countenance of her people to earn Cornwall the unflattering title of "West Barbary." The grind of daily life was alleviated now and again by events such as fairs and feast-days, but it was Methodism which provided both emotional fulfillment and practical aid for the Cornish. The Anglican Church had never been over-strong in Cornwall, and Wesley's teaching - with its emphasis on self-help and individual improvement, and its preoccupation with the next world - met a vital social need, thus eradicating the worst excesses of drunkeness and violence and giving the Cornish a new-found purpose in life.

Methodism also proved an impetus to political development,

7

providing the basis for a peculiarly Cornish band of liberal radicalism which is with us still. But for the Cornishman facing hard times emigration was often a more practical alternative to industrial or political conflict. The Cornish miner (or farmer or fisherman) was confronted with economic tendencies over which he (or often even his master) had little control, and against which it was difficult to combine. Thus emigration became a central theme in nineteenth-century Cornish history. Just as the first three decades of the century had seen the dramatic rise in the populations of various mining parishes, so the remaining decades bore witness to severe depopulation all over Cornwall, from the Tamar to Land's End. The parishes of Breage and Germoe lost 27 per cent of their inhabitants in the ten years between 1841 and 1851, whilst Perranzabuloe lost 22 per cent of its population in the period 1871 to 1881, and St. Cleer lost 25 percent over the same years and a further 22 percent from 1891 to 1901.

In 1881 it was estimated that the mining population of Cornwall had diminished by 24 percent in the preceding decade, and it is clear that many of these migrant Cousin Jacks made their way "Down Under." For years the *West Briton* had carried articles which had asserted that "Large numbers of the mining population are emigrating to Australia..." (October 1863) and that "... Australia and New Zealand... put in their claim for labour, and many hundreds find their way to those colonies..." (June 1866). In September 1875 the same newspaper made the startling announcement that "Ten thousand five hundred and seventy six emigrants... left Cornwall for the Australian colonies during the first six months of the present year," an echo of earlier and almost as incredible statistics, like those of May 1867 which showed that "During the last twelve months, Cornish miners to the number of 7,380 have left the county... 670 (for) Australia and New Zealand." Indeed, in the mid-1860's the "Cornwall Central Relief Committee" was set-up to aid those who wished to emigrate, itself an echo of an earlier company founded in 1840 by the "worthies" of Cornwall (Eliot, Lemon, Molesworth, Pendarves, St. Aubyn, and Vivian) to assist emigation to Australasia. The Cornish miner who remarked to George Henwood in the late 1850s that "I only wish to God I had the money, I'd soon be off to Australia" was not expressing by any means an unusual sentiment.

It was not only to Australia that the miners went, of course. North America was perhaps the greatest home of the Cornish migrant, with South Africa gaining importance towards the end of the century, and with many other areas of the New World and the British Empire taking their quotas of Cousin Jacks. But, that said, the Australian colonies were for many years a principal destination for emigrant Cornishmen, especially miners and their families.

Cornish connections with Australia can be traced to the earliest days of European settlement there. In 1787 Lieutenant Philip King, RN, from Launceston, sailed with the first convicts to Botany Bay in New South Wales.

It was he who surveyed the local coast and recommended the re-siting of the colony at Port Jackson (Sydney Harbour), and he also established the first settlement on Norfolk Island. King became Governor of the island, and in 1800 was appointed Governor of New South Wales itself. Two of the convicts who sailed with King were Mary Broad from Fowey (a petty thief) and William Bryant, a Cornish fisherman whom Mary married in Australia. In 1790 Mary and William, with their two children and seven other convicts, escaped from the colony and sailed the 3000 miles to Timor in a small boat. They were arrested and brought to Britain, William and the children dying during the long voyage. Mary was imprisoned in Newgate, but was freed in 1793, then returning to her native Cornwall.

Equally eventful was the life of Captain William Bligh of "Bounty" fame, who was born in St. Tudy in 1754. Following the well-known mutiny, he returned to Britain and fought under Howe and Nelson in several important actions, and was in 1805 appointed Governor of New South Wales. His thorough-going attitude to his duties and his somewhat rigid approach to discipline earned him various enemies and many were jealous of his position and reputation. In 1808 he was forcibly deposed from his post, and imprisoned until 1810, but he later returned to Britain again where eventually he was promoted Vice Admiral.

Van Diemen's Land (later Tasmania) was first settled in 1803, its early nomenclature reflecting a distinct Cornish influence. Not content with a Launceston, Redruth and a Falmouth, it sported also a County of Cornwall and even had its own River Tamar! However, despite this early Cornish impact in the penal colonies, the real beginnings of the story of the Cornish miner in Australia are to be found in the late 1830s - when a trickle of Cousin Jacks made their way to the then newly-formed colony of South Australia. It was here, only a few years later, that Australia's first metaliferous mineral discoveries were made, and South Australia became thus the first Antipodean settlement to attract the Cornish in any numbers.

Unlike New South Wales, Norfolk Island, and Van Diemen's Land, South Australia was not a penal colony and was instead a remarkably civilised settlement, renowned for its religious and civil liberties. It was colonised according to the systematic and typically-Utilitarian "Wakefield System" and its Liberal Nonconformist flavour earned it the title of "Paradise of Dissent." This no-doubt appealed to Cornishmen, few of whom were immune to the influences of Methodism in Cornwall, and it is significant that one of early South Australia's most effective advocates and propagandists was one John Stephens. He was the son of a Cornish miner-turned-Methodist minister, whose family hailed from the ancient boroughs of Helston and Tregony. His publicity work on behalf of South Australia culminated in 1839 in the publication of his book *The Land of Promise,* in which he argued persuasively the case for migration to the colony, and no-where did his cry have such an intense and long-lasting effect than in the land where his own forebears had lived for generations - Cornwall. It is in South Australia, therefore, that we must begin our search for the Cornish miner "Down Under."

PART 1:

THE CENTRAL COLONY

Chapter 1

THE GREAT MIGRATION

The great mass of Cornish immigration into South Australia occurred between 1836 and 1886, from the time the colony was first proclaimed until the assisted passage scheme was finally at an end. Immigration records are fragmentary and incomplete, but those which have survived show some 12,967 migrants from Cornwall out of a total of 162,853 - i.e. 8 percent. Allowing for missing data and the not-inconsiderable Cornish immigration from other Australian colonies (not shown in records), the total number of Cornish may have been 16,000 or even higher. Population estimates based on this information, together with an analysis of Cornish surnames in South Australia in 1900, have indicated that there were some 30,000 people of Cornish birth or direct descent in the colony at the turn of the century.

The foundation of South Australia, and its parent city of Adelaide, in 1836 coincided with the dawn of the "Great Migration" from Cornwall in the 1830s. Colonisation began with the arrival of the South Australian Company's vessels in July and August of 1836, and the nineteenth-century historian John Blackett recorded that a certain Samuel Stephens was "... the first adult colonist to put foot on South Australian soil." Samuel was a brother of John Stephens, the South Australian propagandist, and thus the son of a Cornish miner. In retrospect, it is rather symbolic that Samuel Stephens should have been the first official settler to land in the colony, for in so doing he established what was to be a fundamental and enduring link between Cornwall and South Australia.

Stephens was joined shortly by literally hundreds of other Cornish migrants - the result of a concerted campaign in Cornwall in the late 1830s and 1840. During 1838 meetings were held in the larger Cornish towns, such as Bodmin, Helston, and St. Austell, with lecturers explaining the benefits of emigration and enthusing over the supposed magnificence of the new colony of South Australia. Local agents - A.B. Duckham in Falmouth, G. Jennings in Penzance, Issac Latimer (a *West Briton* reporter) in Truro, Mr. Geake in Launceston, and J.B. Wilcocks across the border in Plymouth - were appointed during those early years to select suitable migrants and arrange for the sale of South Australian land to intending colonists. These agents held public meetings throughout their respective areas and published informative posters, inviting would-be migrants to call at their offices to be interviewed. Free passages were awarded to those applicants who had suitably impressed the agents and were otherwise qualified to go to South Australia.

Typical of South Australian posters displayed in Cornwall at the time was one circulated in and around Falmouth in June 1839 by A.B. Duckham. Its main text was a long and wordy letter from a Cornish settler in South

Australia, one Marmaduke Laurimer from Falmouth. The account Laurimer gave was seemingly candid, but also subtle and ultimately very persuasive - just the kind of propaganda that Duckham needed - for it began by admitting shortcomings ("... shoes are dear, earthenware very dear; ... Cornish ploughs would be broken to pieces in our soil...") but concluded with fulsome praise ("... the buildings would grace London itself; there is no colony in history has risen so fast as South Australia.") Duckham would have been particularly pleased to publish the section of Laurimer's letter refering to the colony's" ... slate, stone, copper, silver, which in a few years will make it as rich as any country..." Although no serious mining commenced in South Australia until 1841, such references were certain to attract the attention of mining men and create some excitement in Cornwall.

Isaac Latimer, too, made extensive use of letters written home by Cornish colonists as testimonials - one poster issued by him in Truro in October 1839 quoting at extraordinary length from a number of letters. He worked tirelessly and well to promote the colony, on Tuesday 22nd August 1839, for example, delivering a lecture at the King's Head Inn, Chacewater, and on October 15th addressing "... all persons engaged in useful occupations ..." in Bodmin concerning the prospects of emigration in South Australia. The interest such talks generated is evident in the fact that as early as August 1839 the *Royal Cornwall Gazette* newspaper was publishing items of South Australian news. Indeed, between 1836 and 1840 some 941 applications for free passage were lodged in Cornwall - 10 percent of applications in Britain and Ireland. Some 500 of these Cornish applicants were actually accepted which, as whole families were often included on a single application, would correspond to about 1,400 individual persons.

At first, during 1836, it was artizans in far-western Cornwall who displayed the keenest interest in the new colony - men like James Bennetts, a carpenter from St. Levan, and his cousin Pascoe Grenfell, a wheelwright from Madron - but by 1837, when 79 applications were received from adult males and a further 23 from single women, the range of occupations and geographic origins had widened. The major occupational group were labourers - William Batten from Altarnun, James Pedler from Tywardreath, and 24 others - while only six applicants were miners: one from St. Austell, one from St. Blazey, one from Gwennap, and three from Illogan. Most of these 1837 applicants were accepted, many of them voyaging to the new colony in the ship "Red Admiral" - families such as the Sleeps from Linkinhorne and the Williamses from Redruth.

During 1838, when the programme of lectures began to take effect, there were more than 170 applications for free passage to South Australia lodged in Cornwall, from parishes as far-apart as Tresmeer, Perranarworthal, and Towednack, and from individuals as diverse as a roper from Torpoint, a thatcher from Hayle, and a malster from Ruan High Lanes. But this time only two miners - 26 year old Thomas Coon and family from St. Blazey, and Peter

Medland from neighbouring Biscovey. The total absence of applications from the great copper parishes of the west can only suggest that 1838 was a year of relative prosperity and good employment in the mines, there being as yet little sign of the hungry years ahead.

It was not until the following year, 1839, that the miners achieved any kind of prominence - at a time when the total number of applications more than doubled to reach almost 360. Forty-five applicants were miners, but only ten were from Gwennap, the heartland of Cornish copper, the rest being spread through a range of districts such as Camelford, Creed, Kenwyn, Luxulyan, and Perranzabuloe. The stream of Cornish applications continued into 1840, with the miners this time achieving numerical dominance, their 132 applications hardly matched by the 27 agricultural labourers and 12 blacksmiths out of the total of slightly under 300. Although Gwennap could only muster one miner-applicant, the vast majority came from the adjoining districts of Camborne, Illogan, Perranzabuloe, Redruth, and St. Agnes, (with an additional sprinkling from the Tamar Valley at Calstock and Stoke Climsland.) Suddenly that whole mining district, which in 1838 had shown hardly the slightest interest in South Australia, was seething with applicants - Isaac Barkla, miner of Mengoose in St. Agnes; Sukey and Jane Fletcher, bal-maidens from Wheal Burton; Charles Glasson, miner and husbandman of Old Chapel Street, Camborne ... the list seems almost endless.

It is difficult at first sight, perhaps, to account for this sudden exodus from Camborne-Redruth and environs, for in 1840 the copper industry was not yet in trouble, nor had mining operations commenced in South Australia. However, Cornwall in 1840, on the eve of the disasterous "Hungry Forties," was beginning to discern the first painful pangs of a general economic depression, and these would have been felt first of all in the over-populated industrial areas which had expanded so rapidly since the 1780s and were now susceptible to food and other shortages.

In addition, the approach of hard-times would lead the Cornish miner to reflect anew upon his lot in life - the harsh, unpleasant conditions in which he worked (man-skips and man-engines had yet to be introduced!) and the generally unsavoury nature of existence in the mining towns. And in emigrating to South Australia many of these miners were deliberately escaping these poor conditions, and hoped clearly to be able to turn their hands to occupations other than mining. When, once in South Australia, they were drawn to quarrying, well-sinking, and - after 1841 - to mining, it was not the intrinsic delight of digging holes that attracted them, but rather the relatively high wages employers in the colony were prepared to pay for their skills. In 1840 John Paull wrote to his father at Goonvrea, St. Agnes, saying that he could earn from £4 to £5 per week sinking wells, and Thomas Roberts, from Perranarworthal, wrote that miners could earn good money as quarrymen.

That many of these early migrants decided upon South Australia as their destination, which was not at all the "next parish after Land's End" that the United States was, reflected certain of its unique attractions. South Australia was the Liberal-Nonconformist "Paradise of Dissent," and Isaac Latimer could remind Cornish Methodists that

... the vice and demoralization of Australia, has reference only to the penal settlements of New South Wales, Van Diemen's Land, and Norfolk Island ... The morality of the colony of South Australia is secured in every way that can be thought of...

Many believed that, in migrating to South Australia, they were exchanging the land of bondage for the land of the free. Thomas Sleep wrote to his uncle in Falmouth and exclaimed that "... none of us desire to return to the bondage which holds our fellow-countrymen ...," and John Holman informed his father that "... I am freer than when I was in England (sic) ... we would not be back to Southpetherwin for £500." John Oats, a Cornish miner, wrote in even stronger terms, telling his friends at home that to remain in Cornwall was "... to bind yourself in slavery all the days of your life ..."

Letters such as these, written home from South Australia, had the effect of precipitating further emigration from Cornwall. Many were reprinted in local newspapers, such as the *West Briton,* and others found their way into the *South Australian Record* and other journals published in London to promote emigration to the colony. The effect of these letters was so noticable that the *Register,* printed in Adelaide, could remark that

... enough has transpired through the press or the private communications of those who have cordially sent home their favourable impressions to arouse the attention of the enterprising Cornish of all classes ...

Some Cornishmen dwelt upon the material benefits of emigration. Joseph Orchard, from Mawgan-in-Meneage, declared that "... here is plenty of work, plenty of meat, and plenty of money ...," while in 1839 Charles and Mary Dunn, from Trewen, advised their friends to emigrate "... for industrious men are wanted." Thomas Scown, from Launceston, wrote that Cornish migrants were much sought after by employers in the colony - "Londoners in South Australia, are already put by by Cornish men."

This initial burst of migration, and the attendant optimism and euphoria, was cut short at the end of 1840 when South Australia was plunged into financial crisis and economic depression. The initial land boom had spent itself, business confidence fell to a low level, many farmers were almost ruined, and companies could not meet their debts. Capital and labour began to flow from the colony, and Governer Gawler attempted to balance the Budget by adopting a severe deflationary policy during 1841. Just as some

14,000 colonists had arrived in South Australia between 1836 and 1840, some 1,400 being from Cornwall, so the intake was cut back in 1841, with only 145 persons arriving in 1842 and a full flow of colonists not restored until 1844. Those who had the misfortune to arrive at the height of the crisis found South Australia much-altered, one of these unfortunates - James Sawle from Truro - writing that "The prospects of the Colony are getting worse every day ..." and that, with the outbreaks of fever and dysentery, "... I could name many who left Cornwall, who have found a grave in Australia."

Luckily, the strict economic measures taken were sufficient to save the infant colony. A bumper harvest in 1842 showed that South Australia was recovering, and the development of the Glen Osmond silver-lead mines near Adelaide - after the discovery of Wheal Gawler in 1841 by two Cornishmen - seemed to herald the dawning of a new mineral age and increased the clamours for renewed immigration. As a result, the Government began again to finance immigration, and by 1844 streams of colonists were arriving in South Australia under an assisted passage scheme. Needless to say, South Australia's mineral discoveries, firstly of silver-lead, and then of copper at Kapunda in 1843 and at Burra Burra in 1845, rekindled Cornish interest in the colony after the lull of the early "forties." Extracts from South Australian newspapers were reprinted in the Cornish press, and Seymour Tremenheere wrote a paper for the Royal Geological Society of Cornwall on the subject of South Australian mining. Soon Cornwall had her own Burra Burra mine, situated in Kenwyn parish, and a West Kapunda mine was opened at Stoke Climsland in East Cornwall; two further indications of Cornish interest in South Australia.

The Glen Osmond silver-lead mines were said to be "... more than equal to the celebrated East Wheal Rose of Cornwall...," and South Australian papers made great capital out of the startling, almost unbelievable wealth of the mighty Burra Burra. The *South Australian Gazette and Colonial Register* wrote in September 1845 that "The greatest excitement has been produced in Cornwall ..." by the mineral finds. It was proud that the colony was "... a British province, with mines worked by Cornish hands ..." and looked forward to the "... capitalists of Cornwall transferring their energies to the more rich and generous mines of South Australia." The Kapunda and Burra Burra copper mines were soon established as being far and away the richest of all these enterprises, and much of Cornish migration to South Australia in the late 1840s and 1850s was in response to demand from these two workings - especially the latter. By 1845 most ships arriving at Port Adelaide carried contingents of Cornish migrants. The "Isabella Watson," for example, arrived from Plymouth in April 1845 carrying nearly 120 passengers of whom over 30 were Cornish.

The year 1846 witnessed the arrival in South Australia of a number of ships carrying Cornish miners and their families. The "Rajah," for example, brought 30 Cornishmen, a number - such as Uriah Scoble, Peter Spargo, and

John Trenowith - recorded as being engaged to work at the Burra Burra mine. In the June it was reported that another 500 migrants, chiefly Cornish miners and their families, were to leave Plymouth for South Australia during the next few months; and on 5th October the "Kingston" put into Falmouth to pick up 66 passengers, almost all of them miners selected by the South Australian Mining Association - the company which ran the Burra Burra mine. In the November it was noted that passengers on board the "Princess Royal," then lying at Plymouth, were "... chiefly from Cornwall ..." and that South Australia "... owing to its having been colonized chiefly from the West of England, has more of this class of men (Cornish miners) than, perhaps, any colony in our possession..." The passengers on the "Britannia" and the "Hooghly," other arrivals at Port Adelaide during 1846, were also mainly Cornish, being a party of miners under Captain Richard Rodda, from St. Austell, who had been engaged by George Fife Angas to work the various copper deposits on his land in the Barossa Valley, north of Adelaide.

During 1846 specimens of the dazzling Burra Burra ore - beautiful green malachite and deep-blue azurite - arrived at the office of J.B. Wilcocks at the Barbican, Plymouth. The *Devonport Telegraph* estimated that "... the richness of the specimens exceed even those from the far-famed mines of South America ...," and Wilcocks was not slow to use them as part of his propaganda campaign to win recruits from Cornwall for Australia. Wilcocks, indeed, was in the years ahead to play a central role in the selection of Cornish miners for South Australia. He took great pride in his work and was at all time methodical and meticulous. Of the miners he despatched on the "David Malcom" in 1846 he wrote that they were "... as fine a body of people as ever left England." He also commented individually on the migrants, so that John B. Tregea was described as a "Very superior miner," while James Rundle was "A good wheelwright, miner, carpenter and excellent character," and William Spargo was "an excellent captain."

1847 was a repeat performance of 1846, the "Theresa" sailing in January with over 230 emigrants for Port Adelaide, "... most of those on board were from this (i.e. Devon) or the neighbouring county, Cornwall ..., they consist principally of miners and agricultural labourers and female servants." In the October the "China" set sail with a body of miners selected by Wilcocks, and in the following year, 1848, Colonel Carlyon of Tregrehan (a stately home near St. Austell) wrote to Captain Richard Rodda in Adelaide saying that there were "... many still emigrating from this neighbourhood to Australia." At the same time, the *West Briton* noted the departure of many from the East Cornwall villages of St. Cleer and Tremar for South Australia, and the period also witnessed the migration of some 600 people from the St. Just mining area to the colony.

This massive exodus from St. Just-in-Penwith reflected the fact that it was the potato-growing district of far-western Cornwall which was most acutely affected by the potato blights of the 1840s, when local crops were all but destroyed in 1845 and again in 1846. During the hard winter of 1846-47, bands of angry and starving miners beseiged the major towns, and on one occasion a full-scale riot in Helston was only narrowly averted through the distribution of bread to the poor. These "Hungry Forties" were grim days for Cornwall, resulting in further emigration and depopulation. In 1847 William Allen, a migration agent at Penzance in the heart of the potato district, wrote that "Many persons in the Penzance district are preparing to emigrate to South Australia, and amongst them a fair proportion of first-rate miners." In Adelaide, the *Register* further noted that "Mr. Allen says business was dull in Cornwall, and as the potato crop in his area participated in the general failure, much distress was felt and anticipated," and it concluded that this would result in further migration to South Australia for "... Cornish calculators could not fail to draw conclusions highly favourable to the colony ..."

Of course, the *Register* was correct. A contemporary observer, J.R. Leifchild, estimated that during 1849 nearly five percent of the Penzance Poor Law Union had emigrated to Australia and New Zealand, and in the following year, 1850, 50 persons left the parish of Mawgan-in-Meneage on the Lizard to go to South Australia. This movement was reflected in the arrivals at Port Adelaide, so that during 1849 a whole stream of ships arrived in the colony from Plymouth, carrying Cornish migrants. The "William Money," for example, carried 366 passengers of whom half were Cornish, while a similar proportion of the 266 migrants on the "Pakenham" were also from Cornwall. The "Prince Regent" carried 60 Cornish settlers, the "Eliza" 42, the "Himalaya" 53, and so on.

As in the early days, Cornish miners arriving after the revival of immigration in 1844 wrote back to their friends and relatives in Cornwall, telling them what they thought of South Australia. Inevitably, these letters were full of news about the mining boom and, as before, served to stimulate yet further migration to the colony. One miner wrote to a friend in Cornwall, "Oh! Richard, it would make your mouth water to see the Burra Burra mine." Another, John Davey, wrote in 1846 that he was earning £3 to £4 per week in the Burra Burra mine and that "Miners get more wages than they do in Cornwall in two months." Wages were a factor mentioned in many letters written home, for even as late as 1850 the average wage of a miner in Cornwall was only £3 per month. Peter Medland from Biscovey, who was earning 30s to £2 *per week* in the Glen Osmond silver-lead mines in 1847, urged the emigration of "... all the miners of Biscovey and Turnpikegate out here ... in such a flourishing country where there is plenty of everything to nourish and cherish you ..." At Kapunda the story was the same, John Oats writing to his brother in Cornwall:

Here is the place to live! ... if you could but see how we are living you would not stop home a day. The gettings, when we arrived on tutwork, were 10s per day, but I got a great deal more than that on tribute. An industrious man need not work all his days here, for he can get paid well for his labor and live cheap. Me and Caroline can live on 10s per week, and you know that I like a good living.

Any number of similar letters could also be quoted, like that of a Launceston migrant in 1845, or that from Mr. Rendell from Linkinhorne in 1847, or that of John Martin from Stithians who in 1848 thought that South Australia was "... full of mineral of the best samples in the world." Such correspondence was usually successful in persuading kinfolk in Cornwall that migrants were better off in the colony, some settlers employing quite strong language to press their point: in 1846 Samuel Robins wrote to his sister -

Penryn is nothing to Adelaide - we can buy everything we want, from a needle to an anchor, we have schools, chapels, and other institutions that are needful ... You think this place is a wilderness - you are as much mistaken as though you were to say Plymouth is in France. I am determined to stop were (sic) I am ... the remarks in your letter concerning Australia, it makes the heart sick ...

Some, like William Prowse from Penzance, had to overcome their relatives' fear of the voyage - "... you are as safe aboard of a ship, as you are in your own house ... tell Aunt Alice to come and not be afraid of the sea and mother likewise." This fear, however, was not irrational. As the Cornish well knew, wrecks along the coasts of Cornwall were almost commonplace then, and the sheer distance of the journey to Port Adelaide combined with hair-raising tails of conditions on certain emigrant ships to deter not a few would-be colonists. The journal compiled by a Cornishman, one George Richards, lends a grim insight into conditions endured by passengers in the "Java" in 1839. Issac Latimer had written that the "... "Java's" ... accommodations are unusually spacious and lofty, and are so arranged as to ensure the comfort of all the passengers," and an advertisement in the *West Briton* described her as a "... fine first-class teak-built ship."

However, the "Java" was only a few days out of Plymouth when a fatal outbreak of whooping cough occurred, and after a month at sea Richards wrote that "... the intermediate cabins are insufferably hot, full of Cockroaches which destroy the clothes ...," while there was "... beef thrown overboard, pork stinking." Many children died - on one occasion there was the gruesome spectacle of a coffin bursting open as it hit the sea - and Richards recorded the death of the daughter of one of his Cornish friends: "Girl to Bastian of Crowan died aged 11 years." Not long after, his own daughter passed away - "Dear little Caroline died this morning about 5 a.m. committed to the deep 12 o'clock about 4000 miles west of Australia." Even the Neptune jollifications, held to celebrate the crossing of the Equator, degenerated into a violent brawl, and when the "Java" finally dropped anchor at Port Adelaide the passengers were determined to "... expose the shameful conduct of the Doctor, Captain and 3 Officers or Mates towards the Emigrants and Crew."

Richards felt, however, that the principal blame for the conditions on the voyage lay with the Commissioners in London "... for sending so many children 3 to each adult without sufficient quantity of food." Various Passenger Acts - of 1835, -43, and -47 - regulated the number of migrants a ship could carry and determined the appropriate levels of provisioning, but still the *West Briton* in 1849 could note that Cornishmen bound for Australia had to improve life on-board ship themselves by purchasing extra goods before they sailed. When Richard Moyle came out from West Cornwall in the 1840s he found conditions on the ship "disgraceful," while Nicholas Boaden from Veryan complained of "... pigs" slush on deck." The behaviour of some of the migrants, too, added to the misery. In the "Canterbury," bound for Port Adelaide in 1866, there was a squabble when "... a Cornishman named Richards stole some bread ...," and when Francis Treloar journeyed to Australia from Penryn in 1842 he noted the unpleasantness caused by violence amongst the crew members.

Weather conditions added to the trauma of the voyage, Joseph Orchard from Mawgan-in-Meneage writing in 1848 that the ship in which he travelled, the "Westminster," was damaged badly in the Channel and in Biscay, with the migrants "... momentarily expecting to go to the bottom." These dangers were compounded by an outbreak of fever, with Orchard - less than a month out of Plymouth - recording the death of the two-month old son of Henry Lethlean from Camborne. A few days later a baby born to the wife of James Mill of Redruth also died. One year old James Repper from Breage was the next to die, and Henry Samson from Illogan lost his daughter a few days later, while his wife suffered a miscarriage the following day. Tristram Rowland, from Camborne, lost his wife the day before the ship arrived at Port Adelaide.

Certain ships, like the "Lady Ann" in 1859 on which Thomas Treleaven from Boscarne (near Bodmin) was a passenger, arrived at Port Adelaide partially dismasted and in a frightful state of disrepair. Alfred Tresize, a miner from Botallack, was in the Adelaide - bound "Amoor" which was wrecked in Plymouth Sound, and John Liddicoat, another Cornishman, was in the ship "Marion" which was wrecked on South Australia's Yorke Peninsula, in 1840. The emergence of ironclads and steamers made the voyages increasingly safer, of course, and general conditions did improve as the century wore on. But still in 1875 the Rev. R. Carlyon Yeoman, from Perranwell, could note that a stormy passage to Port Adelaide was made all the worse by the sale of intoxicating liquor on the "Collingrove:" "The drinking portion of the passengers were a continual annoyance to the temperate party, especially at night."

20

But despite all this, the Commissioners were concerned for the welfare of the migrants. They constructed a fine Emigrant Depot at Plymouth, where board was 6d a day, and newspaper reporters given the opportunity to inspect the "Princess Royal" prior to its departure for Port Adelaide in 1846 were genuinely impressed by its facilities, writing that the 'tween decks were" ... so constructed as to allow a free circulation of breeze from the windsails, and numerous scuttle admit light and air." J.B. Wilcocks, the migration agent, would always address the migrants on the eve of their departure, stressing that they should ensure absolute cleanliness in the 'tween decks at all times, and that they should avoid arguments and disputes.

The emigrants were themselves usually intensely optimistic as the ship prepared to sail. In the "Princess Royal" in 1846 John Rowe, a Cornish miner from Lanhydrock parish, sang to the assembled migrants a little composition he had written himself. It was a touching, highly allegorical piece, which tells much of the hope and faith experienced by Cornishmen as they prepared to leave for a new home, and a new life, overseas:

> Come on, my brethren, let us sing,
>   Unto that city bright,
> There need not one be left behind,
>   For Christ does all invite,
> And to glory we will sail,
>   We'll sail,
> And to glory we will sail.

It may be that many who shared this optimism quickly lost it when the ships put to sea, but, despite the conditions, many of the emigrants displayed an admirably stoic attitude in the face of adversity. It helped considerably if the doctor was competent and sensitive, and took a genuine interest in the passengers' welfare. Samuel Bray, from Falmouth, wrote that his wife was often ill but that "..so our doctor was very kind, and got her name on the list of those who received one bottle of porter in two days. On the whole, we got on very well." To pass the time the emigrants organised quizzes and ran ship's magazines, a contributor to the "Isabella Watson's" journal in 1845 writing on the Consolidated Mines in Gwennap, which he called "The First Mine in the World." When the master of the ship happened to be Cornish (as many were) it instilled confidence in those passengers from Cornwall. When William Rowe travelled to Port Adelaide on the "Prince Regent" in 1849 he found that the master was one Captain Jago, and when Joseph Hancock set sail for the colony in 1860 in the "Lillies" he found to his delight that the master was none other than his old friend, Captain Williams from Mevagissey. Not surprisingly, there is some evidence to suggest that Cornish sea-captains were inclined to treat Cornish passengers with a certain degree of favouritism!

News of these happier experiences also filtered back to Cornwall, persuading many of the faint-hearted to take the plunge, and it is clear that by the end of the 1840s the Cornish colonists represented a sizeable proportion of the South Australian population and were recognised as such by their fellow settlers. The total population in 1849 was 52,904, with the Cornish portion being in the region of 5,800. Robert Dare, in a letter to his parents in England, wrote that "The principal part that come out here are Cornwall miners ...," and a Wesleyan minister wrote home to Britain requesting further missionaries as congregations were being swollen by newcomers from Cornwall. Even more telling was the September 1849 issue of the *South Australian News* which carried a short-story supposedly reflecting life in the colony, in which the fictional characters were given Cornish surnames such as Trefusis and Vivian.

The relative strength of the Cornish community was also reflected in the emergence of the short-lived but influential "Cornwall and Devon Society," whose prominent members and committee-men were drawn from local mine captains and Adelaide businessmen - men like James Curnow of Grenfell Street and P.G. Moyle of Hindley Street, or Captain Lean of Wheal Prosper and Captain Gundry of North Kapunda. The aims of the society were to ".... watch over the interests of Devon and Cornish colonists ..." and to "... encourage Emigration direct from the Counties ..." It corresponded with mine captains in Cornwall to publicise "... the highly remunerative employment that awaited the skill and enterprise of Cornish miners." In early 1851 the Society petitioned Governer Young, insisting that miners recruited for the colony should be solely from Cornwall and the adjacent mining region of Devon, and that more ships should be dispatched direct from Cornish ports. The Governer replied by stating that between June 1850 and May 1851 one-seventh of all colonists from the United Kingdom were Cornish miners.

The "Cornwall and Devon Society," despite its initial impact, seems - like many other bodies - to have fallen victim to the Victorian Gold Rush, when so many South Australians went off to the neighbouring colony in search of the yellow metal. Although the South Australian economy in general, after the initial dislocation, prospered as a result of the Rush, the copper mines did not. They were faced suddenly with disasterous labour shortages, and each one - including the mighty Burra Burra - was forced to suspend its workings.

The Burra Burra mine - or plain "Burra" as it was known - was then easily the largest and most influential business concern in the colony, and its owners, the South Australian Mining Association (SAMA), reacted to the labour shortages by attempting to recruit further supplies of miners direct from Cornwall. Henry Ayers, the SAMA Secretary, began to apply pressure during 1850 and throughout 1850 and 1851 J.B. Wilcocks sent out streams of Cornish colonists. For example, the "Stag," which arrived at Port Adelaide in June 1850, carried 32 migrants from Cornwall - some, like John Grigg and Joseph Hooper, having been engaged specifically to work in the Burra. The

"Ascedant," which arrived in January 1851, carried 50 Cornish passengers, the "Sultana" 59, the "Omega" 131, and so on. And when the news of the Victorian Rush led to the wholesale departure of Cornish miners from the colony in 1851 and 1852, Ayers realised that this already considerable flow from Cornwall would have to be even further increased if the South Australian mines were to stand any chance of maintaining production.

Accordingly, in November 1852, Ayers wrote to J.B. Wilcocks, praising him for his "... judicious selection of Cornish and Devonshire Miners for this colony ..." but adding that "We could find employment for a thousand hands consisting of Miners, Smiths, Engineers, Carpenters and others employed at Copper Mines - at wages varying from £6 to £10 per month." But the state of the South Australian labour market continued to deteriorate, and only a month later Ayers was again writing to Wilcocks, protesting that "We are thirsting for labour ..." At first Wilcocks had difficulty in responding to Ayers' demands, for Cornish attention was now fixed firmly on Victoria, but nevertheless ships arriving during 1852 and 1853 continued to sport large Cornish contingents. By 1854, however, the labour situation had not improved significantly, and so this time Henry Ayers offered J.B. Wilcocks a personal bounty of £2 per head for up to 500 Cornish miners. The SAMA Directors also sent a further £500, "... to be disbursed ... in small sums among intending Emigrants from the Mining population of Cornwall ..."

Wilcocks responded magnificently to Ayers' gesture, sending over 1600 Cornish migrants, mainly miners and their families, to South Australia during 1854 and 1855. Further sums to aid Wilcocks with his "bribery" were sent, and he even received the Adelaide *Observer* to keep him informed of local events. Although Ayers was not entirely happy with the miners who arrived on the "William Prowse," the first vessel to arrive after the bounty offer, because he found them too young and inexperienced, he was delighted by those on later ships - those on the "Nile" being "very superior" men who "... will meet with immediate employment."

Towards the end of 1855 the Victorian gold-fields experienced a new wave of popularity, and Ayers had yet again to write to Wilcocks to ask him to increase the rate of selection in Cornwall. He also wrote to the Colonial Secretary in Adelaide railing against the increasing practice of despatching colonists from Southampton, instead of Plymouth, for "... the cost of joining a ship at Southampton, is far greater than the majority of Emigrants from Devon and Cornwall can afford to pay." Ayers was also furious about the way in which some migrants abused the assisted passage scheme to South Australia by making their way to Victoria as soon as they had landed at Port Adelaide. He forwarded the names of offenders (many are recognisable as Cornish) to the Colonial Secretary, with the implication that some action ought to be taken, but in practice it was almost impossible to prevent the abuses. As early as June 1852 migrants had had to sign a declaration to the effect that they would not go to Victoria, but his did not deter the most ardent gold-seekers, and even in 1857 it was still necessary for Wilcocks to interview all migrating miners to ensure that they were not going to cynically take advantage of South Australia's liberal scheme.

In his attempt to secure a steady flow of Cornish miners to the colony, Henry Ayers also made use of the "nomination" system which was experimented with from time to time in the 1850s, inviting miners at the Burra to nominate friends and relatives in Cornwall. In certain instances the SAMA was prepared to pay the full amount to secure the migration of miners, but there were those who took a dim view of this discriminatory practice, asking if it was "... becoming to make this offer of assistance to *Cornish* alone?" Ayers, in characteristic fashion, continued to ignore such criticisms.

In March 1856, for example, he authorised J.B. Wilcocks to send out - at the SAMA's expense - the wife of William Kent (the chief timberman at the Burra) who was then resident at Probus. On other occasions, when those nominated were too old to qualify for a completely free passage, the SAMA undertook to pay the remaining part of the fee. In 1850, for instance, William Richards of Herodsfoot and Mary Goldsworthy of Tywardreath were both over 50 years of age, and so SAMA was required to pay £11 each towards their passages. As it happened, Richards died on the eve of his departure from Cornwall, and soon Ayers was writing to the Colonial Secretary asking for his money back!

The "nomination" system, despite certain advantages, had a major disadvantage in that it was not easy to discern between "good" and "bad" nominees. J.B. Wilcocks, as already noted, took great care and pride in the selection of his migrants, writing of those in the "Hooghly" in April 1856, for example, that they were "... from Cornwall and the mining districts of Devon - they have been selected with much care I hope there is not an inferior labourer among them." However, he was irritated when the "nomination" scheme was again tried in 1857, because migrants were chosen by nomination rather than by selection. Of the miners who sailed on the "Lady Ann" he said that "... a more miserable lot of people never before left the shores of England," and Ayers was forced to agree that "... you (Wilcocks) would have been severely censured ... had you selected such people." But by 1858 the debate was becoming increasingly academic, because an anti-immigration campaign in the colony, spurred on by a partial failure of the wheat and wool crops, was beginning to have an effect. In 1859 only six shiploads of migrants arrived in South Australia, and by 1861 immigration had ceased, albeit only temporarily.

Although the 1850s as a whole, with the effects of the Victorian Rush, had been a damaging era for South Australia's mines, at least there had been a ready supply of miners from whom to draw new recruits. For by the mid-50s the copper mines in West Cornwall - in the Camborne-Redruth district, and especially in Gwennap - had fallen into decline, and although some miners

24

found work in the recently-discovered and still flourishing East Cornwall and Tamar Valley mines, others sought employment overseas. Dolcoath mine had turned from copper to tin as early as 1834, the far-famed Tresavean was stopped in 1853, and the mighty Consolidated Mines were abandoned in 1857. The year 1856 saw Cornish copper reach its highest-ever output - 13,247 tons of metal from 209,000 tons of ore raised - but the statistics belied the real situation, the production figures being boosted by the Caradon Mines (discovered in 1837) and Devon Great Consols (discovered in 1844).

It is interesting to reflect that in the period 1851 to 1860 some 48,886 migrants were recorded in the shipping passenger lists, of whom 5,135 - or 10.5 percent - were from Cornwall. Also of interest is that the vast majority of these Cornish people were copper miners and their families. Between 1848 and 1860 the occupations of a great many adult male Cornish migrants were recorded. Of these 2,117 Cornishmen no less than 1,797 - or 84.9 percent -were miners, whilst many of the others were in allied or service industries such as engineering, blacksmithing, quarrying, carpentry, and masonry. South Australia had truly become one of the great homes of the Cornish miner overseas.

The rapid decline of the West Cornish copper mines was due to their becoming extremely deep, less rich (some almost worked-out), technologically obsolete, and managerially inefficient. On the world market fluctuating copper prices contributed to the erosion of Cornish profitability, while increasing overseas competition gave momentum to the Cornish decline. This competition came first of all from South America, but was soon reinforced by the discoveries of copper in the United States and in South Australia. For example, in 1845 Cornish ore could only make an average of £5 15s 6d per ton, while Kapunda ore was realising £14 15s 3d! Thus it was that the South Australian mines - which depended so much on Cornish labour - were in part responsible for undermining the Cornish mining economy.

When Francis Dutton was in London in 1845 to promote his Kapunda mine, his view that South Australia could come to challenge the Cornish industry was greeted with contempt. "Pooh! Pooh! my dear sir," they all said, "... all the ore you will ever produce from South Australia will be but a drop in a bucket of water!" Dutton's only comment was "Time will show ..." As early as July 1845 the *South Australian Gazette and Colonial Register* had predicted that, with the rise of the colonial mines, "... many of the Cornwall mines ... must be abandoned," while in 1846 the *Devonport Telegraph* feared that "... it is certain that the very rich ores from South Australia will operate most prejudicially upon the mining interests of Cornwall..." As Seymour Tremenheere explained to the Royal Geological Society of Cornwall, South Australia was a wool-exporting colony, with the result that

... the wool-ships on account of the lightness of the cargo are
obliged to take in a large quantity of ballast, and they are

25

therefore glad to take the lead and copper ore at a merely nominal rate of freight; - at the time in question about eight or ten shillings per ton. That circumstance was considered as bringing their mines, as it were, actually into Europe, or at all events as placing them upon equal footing with European mines ...

As South Australian copper increased in output, the *Register* noted grimly that, with this increase being due to the efforts of Cornish miners in the colony, it was now "... Cornwall against Cornwall." It was clear that a vicious circle had become established, with the expansion of the South Australian mining industry further eroding the Cornish economy, and thus creating a still greater supply of unemployed miners anxious to find work in the colony. Their arrival facilitated even further expansion of mineral output, and thus the circle was perpetuated. A brief respite was provided by the Victorian Gold Rush, when J.R. Leifchild wrote that "The scales may be said to be suspended over Australia and Cornwall, and the fall in one produces a corresponding rise in the other." But the vicious circle was set in motion again by the resurgance of the South Australian copper industry in the late 1850s. More particularly, it received new vigour from the discovery of the incredibly rich copper deposits on northern Yorke Peninsula in 1859-61, in a district soon to be known across the world as "Australia's Little Cornwall." As the Kapunda *Northern Star* put it, South Australia was by now "... out-Cornwalling Cornwall altogether."

The development of the Yorke Peninsula mines, together with the excellent harvest of 1861, stimulated demands for a resumption of immigration. One petition circulated at Kapunda called for a "thousand miners" for the colony, and at the end of 1862 the Sutherland Act allocated one-third of the annual land sales revenue to the immigration fund, to finance assisted passages. This Act, coming as it did at a time of deepening crisis in the Cornish copper industry, precipitated a further wave of Cornish migration to South Australia. Throughout the early 1860s Cornish copper mines were closing, while others were desperately trying for tin, and many Cousin Jacks found Australia a far more suitable destination than the United States - for the latter was in the throes of a bloody and damaging Civil War.

In the entire period 1862 to 1870 some 13,265 immigrants were recorded of whom 3,235 - or 24.4 percent - were from Cornwall. And of the 1,007 adult male Cornish migrants for whom occupations were recorded in the period 1862-67, some 366 - or 36.3 percent - were miners, while 317 were labourers (many no doubt engaged in the mining industry), 57 blacksmiths, and 65 agricultural labourers. The great influx from Cornwall started in 1863; and in 1864, a year of heavy immigration, over 16 percent of all migrants were from Cornwall, the West Cornish mining district of Marazion sending as many as 150 people to South Australia. And in 1865, another year of heavy immigration, an incredible 42.5 percent of persons entering the colony were from Cornwall. Every ship arriving that year sported a massive Cornish contingent, 211 out of 315 passengers on the "Queen Bee," 214 out of 358 on the "Lady Milton," 242 out of 388 on the "Gosforth," and so on. (Many of the other migrants on these ships were copper miners from Devon; quite a number were probably Cornish-born).

The Moonta and Wallaroo companies, the two major concerns on northern Yorke Peninsula, took full advantage of the poor economic conditions in Cornwall. The average wage rates they could offer fluctuated between 36 shillings and 45 shillings per week, depending on the price of copper, which were not as startling as those once offered at the Burra but nevertheless more than an employed miner (let alone his unemployed counterpart) could earn in Cornwall. The growing demand for miners at Moonta and Wallaroo could be only partially satisfied by recruitment from Kapunda, Burra, and the Victorian diggings, and so in April 1864 the Moonta Mining Company instructed its agents in London to select miners in Cornwall. It explained that labour shortages "... have induced the Directors to send to Cornwall through you for fifty men to be brought out under the Assisted Passage Regulations...," and shortly after asked the agents for a further 200 Cornish miners.

Demand at Moonta, however, continued to outstrip supply, and towards the end of 1864 T.F. McCoull (the company Secretary) was complaining to Captain Henry Richard Hancock (the mine manager) that ship-loads of miners arriving at Port Adelaide were being snapped-up by agents from the Burra Burra mine. For the widespread departure of men from the Burra to go to northern Yorke Peninsula and to newly-discovered mines in the Far-North of the colony, had created acute labour shortages there, too. As usual Ayers wrote to J.B. Wilcocks for help, and in 1864 he complained to the Colonial Secretary that "... the supply of miners arriving in this Colony is very far from being equal to demand." The Wallaroo Mining Company, too, took steps to recruit direct from Cornwall, and by the end of 1864 a party of its men had arrived in the "Tarquin," while a further supply of miners and smelters were preparing to leave for South Australia.

The great crash in Cornish copper occurred in 1866, and in the ensuing years many of the famous Cornish mines - Great Wheal Busy, Fowey Consols, St. Day United, and scores more - were abandoned, while by 1870 every one of the great Gwennap mines was closed. Although a restrictive immigration policy served to prevent an unchecked flow of Cornishmen into the colony between 1867 and 1871, thousands did leave Cornwall for Australia during those years, and not a few found their way to Wallaroo and Moonta. When a more liberal immigration act became law in 1872, Cousin Jacks again came to the colony in considerable numbers, so that between 1872 and 1879 approximately 7.0 percent of the 24,339 migrants recorded were Cornish.

Again, many of those migrants were miners. In the early 1870s there was still a considerable demand for them on northern Yorke Peninsula, and the Burra, too, in a desperate bid to boost its failing production, endeavoured to obtain further supplies of miners from Cornwall. In early 1874 a group of Cousin Jacks from the "Dalhousie" and "West Australian" were engaged by SAMA, and in December 1875 Captain Sanders - by then the mine's manager - wrote to his brother in Cornwall, asking him "... to make a selection of good Cornish Miners... " But by then the declining mine could no longer afford to pay its men high wages, and miners were leaving to earn more sinking wells on new settlements in the Far North. Sanders was reduced to accosting men as they disembarked from the emigrant ships, in July 1876 it being noted that he had "... secured nine Cornish miners from the late arrival in Adelaide..."

As before, there was an adequate supply of miners from Cornwall. But now, following the demise of copper, it was a crisis in the tin mining industry which was the cause of the migration. During the 1850s and 1860s many Cornish mines had turned their attention to tin, which in 1872 appeared to be the saviour of the Cornish economy, but even by January 1875 the *West Briton* was lamenting that "The year 1874 will be as memorable for the depression in Cornish mining as its predecessor was for its exciting prosperity." Ironically, it was again Australian competition - this time from Tasmanian tin - which had helped precipitate the crisis. The 1870s, too, were a time of agricultural depression in Britain, and although Cornwall was less badly hit than other parts, the depression - coinciding as it did with the tin crisis - caused a parallel migration of Cornish farmers and agricultural labourers. Many went to South Australia where the Government, under the Premiership of James Penn Boucaut (himself a Cornishman), was pursuing a vigorous policy of opening-up the Far-North for agricultural purposes.

By the late 1870s, growing unemployment in the northern Yorke Peninsula mines served to discourage Cornish immigration into the colony. Newcomers wrote home with the news that the great days of South Australian mining appeared to be over, and the local Miners' Union - which had emerged during the 1870s - became increasingly hostile to immigration. The Unionists sent a deputation to the Parliament in Adelaide to call for an end to immigration, and in May 1874 they were amazed to learn that a Mr. Gowling, a newly-appointed migration agent, had decided to pay particular attention to Cornwall when it was already clear that unemployment on Yorke Peninsula was growing. Many felt that their kinfolk at home were being tricked into coming to the colony, when they would be better off elsewhere. A highly satirical article appeared in the *Yorke's Peninsula Advertiser*, purporting to be an account of a typical emigration meeting in Cornwall, in which the supposedly unscrupulous tactics of the agents were exposed. The article, entitled "Report of an Emigration meeting held in St. Just, Cornwall near England ...," gave a selection of "questions" and "answers" designed to reveal both the dishonesty of the agent and the gullibility of the Cornish:

28

Mr. TREVORROW would ask if there were any snakes there?

Mr. GOWLING would answer that there had been snakes in the colony, but the last one had committed suicide several years ago.

Mr. GILES would ask if people stood on their heads out there?

Mr. GOWLING replied that they did sometimes by reason of the earth being spherical. When it was twelve o'clock in St. Just, the earth was upside down in Australia. But of course every honest man would be in bed at that time, so that it was all in favour of the colony, as rogues could not do much injury to lives or property if they had to do it on their heads.

Similarly, in November 1876 Edward Trewidden of Kadina, the township adjoining the Wallaroo Mines, wrote to the *Yorke's Peninsula Advertiser,* explaining how he and his wife had been deceived by the migration agents. He wrote,

What sort of a joke do you call this? I had just got back to Camborne from America where I had been working and picked up a few dollars to marry my little maid, and meaning to take her back to the Lakes, where I had done considerably well. Well, Grace said to me, "Isn't it very cold out there Ned?" "Well, Grace, says I, "it is tarnation cold I must say." "There was a man giving a lecture on going to South Australia, Adelaide," says Grace, "and I thought he gave a pretty good account of it, and it is never cold out there, and I think I could bear the heat better than the cold of America. He said miners' wages was 45s and 50s a week, and that bread and beef, and mutton, and tea, and sugar, were very cheap, and everything else in proportion."

Swayed by Grace's enthusiasm, the couple made their way, after all, to Kadina. But on their arrival they found that "You've come in a poor time Ned ... they're sacking men instead of putting them on ..." and that food was "... as dear as snuff." Others shared Trewidden's disillusion, one newcomer writing in 1877 that, although he found Adelaide "... a fine little town, very little inferior to Truro at home ...," the rest of the colony was ... "one of the most barren places I have ever seen ... Is this the beautiful country we heard so much about at home, the land flowing with milk and honey?" Another blamed the Rev. John Thorne, a Bible Christian minister from Moonta who had done much to publicise the colony in Cornwall, saying that he had heard

"... a new chum cursing that Bryanite preacher called Thorne, as he had said there was plenty of work in South Australia for miners..."

By now the increasingly organised and vocal "Moonta Miners' Association" was entirely convinced that continued immigration mitigated against the best interests of both the Yorke Peninsula men and the New Chums coming out from Cornwall. In September 1879 John Prisk, the miners' leader, argued that the men ought to be "... writing home to their friends in Cornwall ... to give them a truthful account of how things are here." The Association continued to lobby the Government, and it is significant that when finally immigration was severely cut-back, in 1879 and 1880, it was largely in response to public pressure.

A more liberal immigration policy was again instituted in 1881, and this coincided with an exciting - if rather temporary - revival in the South Australian copper industry. This revival was centered principally on the Wallaroo Mines, and as early as May 1881 Captain Hancock was taking measures to aquire new supplies of miners. In October 1882 the Wallaroo Mines Directors decided to send Captain Richard Piper over to Cornwall to act as their recruiting agent, and not long after the Moonta company announced also its intention of procuring 100 Cornish miners. On his arrival in Cornwall, Captain Piper inserted an advertisement in the *West Briton,* which called for "... Fifty good Miners, including steady young men, also married men with their families, to proceed to the Wallaroo Copper Mines..." In early April the Wallaroo company instructed Piper to increase the recruitment number to 250, and at the mines themselves the erection of new company-built cottages was evidence that something was afoot. The locality was alive with rumour - was Captain Piper "... going to bring coolies or Cornishmen? And suppose the Dolcoath Mine stops working will he engage all the miners in a mass to come out here?"

Piper arrived finally in South Australia with 408 "New Chums," most moving into the new cottages at Wallaroo Mines, with 70 going to live at Moonta Mines. Captain Hancock took pains to introduce the New Chums, saying that one had declared that "... he had eaten more meat in Australia in the short time he had been here than he would have eaten in a year in Cornwall," and the recruits were generally well-received by the work-force. But this did not make for universal harmony, for some newcomers felt they had been tricked by Captain Piper, who had apparently described Yorke Peninsula in over-glowing terms. One asserted that Piper had told him that "... the Company had built them five-storied houses, such as adorn the streets of Plymouth...," and others - like John Kinsman, William Annear, and 27 more besides - registered their protests by disappearing from the district before their contracts with the company had expired. News of this debacle, filtering back to Cornwall, would hardly encourage others go to South Australia. And, although the occasional miner still made his way to the colony during 1884 and 1885, attention was by now fixed firmly upon the gold and diamond fields of South Africa.

In 1886 assisted immigration came finally to and end, and the flow of Cornish colonists - which had run so strong since 1836 - dwindled to a halt. There were a few who came out in later years at their own expense, like Albert Nicholls from Gunnislake in 1887, but these were not in any sense a part of the mainstream Cornish immigration. The northern Yorke Peninsula mining district, the remaining focus of Cornish influence in South Australia in the early twentieth century, experienced a second era of great prosperity between 1898 and 1919. But now the demand for miners was met by local supply. In 1923 it was rumoured that unemployed Cornish tin miners were considering emigration to the Peninsula, for "People in Cornwall think Australia is a country of money. Government officials tell them they can earn lots of money mining or farming out there." But it was never a practical plan, or one likely to materialise, for by 1923 the Wallaroo and Moonta Mines were themselves being abandoned.

To all intents and purposes, then, the real migration from Cornwall to South Australia had ceased in 1886. In 1893 a visitor to Cornwall wrote that "... the Cornish miners have mostly emigrated ..." and that, to find the Cornishman engaged in his traditional pursuit, one had to travel to Australia where "... in some Wooloomooloo, or other place of name infinitely repetative, you shall who seek, find him ..." For those who had remained in Cornwall it was time to contemplate the dereliction and depopulation, and reflect upon more than half a century of frantic emigration.

Chapter 2

THE MINERAL KINGDOM

Almost every study of the Cornish miner records, with rather boring regularity, the old adage which asserts that "Wherever in the world a hole is sunk in the ground, you will be sure to find a Cousin Jack at the bottom of it, searching for metal." But nowhere outside Cornwall did the saying ring so true as in South Australia, where Cornishmen played the central role in the development of mining throughout the last century. They were involved in the industry as prospectors, miners, captains, engineers, mechanics, and general labourers, but their influence ran deeper to affect such things as mining architecture, mining technology, and even the organisation and practices of the industry.

Physically, the mines looked like direct transplantations from Cornwall - with their stone-built chimney stacks capped with sections of red brick, their Cornish beam-engines and engine-houses, their horse-whims and count-houses, and even the dry-stone "hedges" constructed in the time-honoured Cornish fashion. Employment in the mines was according to the peculiarly Cornish "tribute" and "tutwork" arrangement, and many mines were run on the Cornish "cost-book" method, while an independent legal system to adminster the mines - in some respects similar to the Stannaries in Cornwall - was set up in South Australia.

The language of South Australian mining was almost entirely Cornish-inspired. Wheal Friendship, Wheal Prosper, and Wheal Virgin were mine names entirely reminiscent of Cornwall, and as early as June 1848 the *West Briton* had occasion to note the widespread use of the prefix "wheal" in the colony. The description "bal" was also used, and colonial mine names such as Callington, Crinnis, Ding Dong, Phoenix, Tresevean (sic), Great Devon Consols, United Hills, Botallack (sic), Carn Brea, Trengoff, Trevue, Truro, New Cornwall, Old Cornwall, and Duke of Cornwall, were drawn direct from Cornwall. Cornishmen sometimes gave their own names to mines they had discovered, so across South Australia one encountered Benalack's Mine, Jenkin's Claim, Kirkeek's Treasure, Trenowden's Claim, Davey's Mine, Paull's Consolidated, and countless more.

Cornish terminology also extended to mining practices, mining implements, and geological formations. A "vugg" was an underground cavity, a "costeen" an exploratory trench, a "stull" a supporting timber, and pyrites "mudic." "Crowst" and "crib" were two words to describe a lunch break, and the term "Cornish" itself was used in inumerable situations - for instance, Cornish engines, Cornish captains, Cornish boilers, and Cornish stamps.

At the economic level, the Cornish mining influence was crucial to the well-being of the colony. It was the discovery and exploitation of mineral deposits in the 1840s - made possible by Cornish enterprise and labour - that rescued South Australia from the economic doldrums, and the mighty Burra Burra attracted much-needed capital and labour, while the wages it paid generated - through the multiplier process - considerable wealth throughout the colony. The same was true of the Moonta-Wallaroo complex, which developed rapidly in the 1860s and remained South Australia's largest industrial producer until 1923.

With one in ten of the early colonists hailing from Cornwall, it was inevitable that Cornishmen wandering through the Adelaide Hills would encounter the many, rocky, mineralized outcrops to be found there - even those entirely unversed in minerology would not have failed to see the startling blue and green carbonates of copper, while the silver-lead gleam of galena would attract all but the most unobservant of eyes. That South Australia was the last Australian colony to be settled but the first to develop its mineral wealth is an oft-quoted paradox. Its explanation lies in the fact that there was considerable prospecting during the economic crisis of the early 1840s, the accessability of ore in the Adelaide Hills facilitating its discovery, extraction, and transportation.

As early as 1836 men talked of finding mineral wealth, but it was in 1838 that James Nicholls - a Cornishman - collected a large number of silver-lead samples in the Hills and traced the lode for a mile, this presumably being at Glen Osmond where "The first undoubted indication of the existence of silver-lead ore was made in 1838, on a section belonging to Mr. Osmond Gilles, at the foot of the Hills near Adelaide..." At the time no efforts were made to develop the find, although with the onset of financial troubles a year or two later, attention was again turned to the possibility of extracting mineral wealth from the colony. It was in February 1841 that Hutchins and Thomas, "... we believe two emigrants from Cornwall..." as one newspaper put it, made a new discovery of silver-lead at Glen Osmond. A small company with a capital of £6,000 was formed and by April 1841 the very first metaliferous mine in Australia, Wheal Gawler, was in production. Soon other mines were at work in the area - Wheal Watkins, the Glen Osmond Union, the Enterprise, MacFarlane's Mine, and Wheal Hardy - with most of the workforce being Cornish. There was Captain J.B. Pascoe, Captain Stephen Lean, George Prout from Camborne and Stephen Carthew from Redruth, and others with surnames such as Trewartha, Knuckey, Thomas, Veale, Ede, Mitchell, Trethowan, and Odgers.

In 1843 came Australia's very first discovery of copper, at Kapunda, some 50 miles north of Adelaide in what was then isolated pastoral country. Francis Dutton wrote that, when rounding-up stray sheep, he came across "... beautiful green moss ..." which on closer inspection proved to be malachite. He confided his find to a neighbouring pastoralist, Captain Charles Bagot, who when admitted that his son had made a similar discovery at the end of

1842. Together the two men aquired the mineralized land, and set about working the deposits. A sample sent to Swansea assayed at 21.5 percent copper and sold for £21 per ton. Suitably encouraged, Bagot "... agreed with Robert Nicholls, a Cornish miner, for a twelve month to work on tribute." The mine was developed in two sections, Wheal Dutton and Wheal Charles, Francis Dutton noting that "... Cornish miners who were quietly following pastoral and agricultural pursuits ..." in the colony were not slow to make their way to Kapunda to resume their old trade, attracted by "... a liberal tribute for the first year (3s 6d per £) to set the thing going."

Ore raised at Kapunda during 1844 and 1845 realised a total of £6,225, and by 1846 a considerable community had grown up around the mine, its flavour being overwhelmingly Cornish. William Trevena from Redruth was the first person to built a house in the district, and he was joined shortly by many of his compatriots, such as William Christopher from Stoke Climsland, Ralph Kestel from Portreath, Henry Waters from Penzance, Thomas Axford from Chacewater, Henry Bastian from Truro, and Henry James Truscott from St. Austell, many living in the adjoining township named Helston! With its large complement of Cornish miners, development work at Kapunda proceeded quickly, but heavy influxes of water during 1847 and 1848 hampered operations. Accordingly, in 1847 a second-hand 30 inch beam-engine was ordered from Cornwall, and it arrived at the end of 1848 - certainly the first mining engine to be acquired anywhere in Australia. The aquisition of second-hand engines from Cornwall was a practice advocated by the *Register,* and it urged other South Australian mines to follow Kapunda's example. Certainly, Kapunda's engine worked wonders, for after its arrival production increased to about 2,500 tons per annum, this level being maintained until the miners left for the Victorian Gold Rush in 1851-52.

The discovery of the Kapunda mines led to other finds in the area - there were soon a North Kapunda, a South Kapunda, and a Wheal Gundry - and it turned the attention of the Adelaide Hills prospectors from silver-lead to copper. The Montacute Mine, only 10 miles north-east of Adelaide, was in operation by early 1844 when 32 tons of copper were extracted under the watchful Cornish eye of Captain Tyrell, and thereafter it was worked intermittently into the mid-1850s by Cousin Jacks like Captain Morcom, William Gummow from Padstow (in Cornwall he had driven beam-engines at Wheal Rock and Perran Great St. George), Thomas Roberts from Perranwell, and a J. Harris from Liskeard.

Adjoining the Montacute were the North Montacute, the Mukurta, and the claims of the Adelaide Mining Company, Francis Dutton considering that the development of all these mines "... will undoubtably be furthered by their engaging, as soon as possible, the assistance of practical Cornish mining captains ..." During 1844 and 1845 the London *Mining Journal* noted the discovery of other mines in the colony. But, after Kapunda, the major copper find was at Burra Burra in virgin bush country almost 100 miles north of

Adelaide. In June 1845 two shepherds made separate discoveries on the banks of the Burra Burra Creek, and both traded the secret of the locations to commercial groups in Adelaide. These rival groups, the "Nobs" and the "Snobs," competed for the land but - to raise the £20,000 necessary for the 20,000 acre survey insisted upon by Governer Grey they were forced to combine their efforts. Having purchased the property, the two groups divided the land - the "Snobs" (the South Australian Mining Association - SAMA) drawing the section to become the Burra Burra mine, and the "Nobs" gaining what was to be the Princess Royal.

The Princess Royal was initially very successful. It was worked during 1847 and 1848 by miners on tribute, under the management of two Cornish captains, Messrs. Bath and Richards, but by 1851 the mine's capital was all but expended. The workings were abandoned and the land sold as a sheep run. The "Nobs," then, had drawn the poor section of the 20,000 acres, leaving the spectacular riches of the Burra Burra to the rival "Snobs." On 24 September 1845 Thomas Roberts - formerly of the Montacute - was sent in charge of nine other Cornish miners to commence work at the Burra. The initial working they called Great Wheal Grey (after the Governer), and the incredible quality of the ore they produced (it assayed at 71.25 percent copper) made it abundantly clear that this mine - the Monster, as it was popularly known - was the richest find in the colony to date. A contemporary verse conveys something of the heady atmosphere which pervaded South Australia following the Burra discovery:

> Have you not heard of the Monster Mine?
> There's never a man to be got to dine,
> There's never a clerk who will pen a line,
> At my behest or thine.
> They are all gone to the houseless North
> To gaze on the Monster Mine.

The captains engaged to run the Burra were almost exclusively Cornish. Henry Roach, from Redruth, had worked in Cornwall at the Tresavean mine and had visited Columbia before coming to South Australia to become General Superintendent of the Burra Burra mine in 1847, a position he was to hold until his retirement in 1868. A visitor to the Burra in April 1848 wrote that he encountered "A Captain Tre-something, and a Captain Pen-something ...," for Roach was careful to recommend for employment as his assistants only men with a thorough knowledge of Cornish mining practice. In June 1847 Matthew Bryant was engaged as second captain, in March 1848 Richard Goldsworthy - from Bodmin - became third captain, and William Mitchell was appointed to the fourth position in 1849. Samuel Osborne became chief ore-dresser, S. Penglaze was "grass captain," Philip Santo - from Saltash - was Clerk of the Works, and John Congdon (formerly of the Caradon Mines in East Cornwall) became Chief Engineer. A Mr. Boswarva was put in charge of clerical and administrative work at the mine, and two Cornish masons, Ambrose Harris and Thomas Paynter, were entrusted with the physical construction of the engine-houses. The actual miners were also mainly Cornish - men such as William Phillips from Gwennap (he had worked as an engine-driver at United Mines), Nicholas Bennett from Constantine (his father had been a captain at the Polberrow mine in St. Agnes) and John Quintrall from Polladras Downs in the parish of Breage.

A clue to the exact proportion of Cousin Jacks amongst the Burra Burra miners is provided by entries in the copper ore day books, wages returns, and similar documents. In 1848, of 63 tributers listed 47 bore either exclusively or typically Cornish surnames. In 1860-61, of the 198 tributers listed 135 had Cornish names - William Trevorrow, Henry Besanco, Thomas Andrewartha, Thomas Yelland, James Tremewan and many more. Only three had German names, while many of those with English surnames such as Hill, Bishop, and Buckingham could well have been Cornish. It was not only the tributers who were Cornishmen, of course: some, like Richard Roscrow, were general labourers, while others - like Ludgvan-born James Thomas and Henry Pinch from St. Mabyn - worked as teamsters.

The Cornish, then, were an all-pervading influence at the Burra mine, and, as at Kapunda, their efforts facilitated rapid development work. In the first six years, 80,000 tons of ore were shipped to the UK, yielding a profit of £438,552, with the Cornish firm of John Williams & Son arranging for its sale at the Swansea ticketings. A number of settlements sprang up at the Burra, and in April 1846 the *Register* declared that "... we shortly expect to find our mining villages vieing with those of Cornwall and Devon." The SAMA township was named Kooringa, but three of the Burra's several other suburbs were graced with the names of Cornish towns - Redruth, Copperhouse, and Lostwithiel. The miners frequented the "Cornish Arms" and the "Redruth Arms," and at Redruth lived in streets whose names were taken directly from home: St. Day, Mevagissey, Tregony, Morvah, Sancreed, Crowan, and several more.

Transport was a problem during the winter months when the Burra road was "... literally strewed with laden drays bogged up to their axles," but the first major difficulty to be encountered was the incursion of water into the deeper workings. Thus, in August 1847, Henry Ayers - the SAMA Secretary - ordered a 50 inch pump engine and accessories, together with a stamps engine, from the Perran Foundry in Cornwall. Not long after, in early 1849, Ayers ordered a 30 inch whim engine from the same source, and four months later he sent also for a second stamps engine of 30 inches. During 1849 and the early months of 1850 it became apparent that, with the mine becoming ever deeper, an even larger pump engine would be required. Thus yet another engine, this time of a massive 80 inches, was ordered from Perran Foundry.

Although Ayers was fearful that the winter rains would inundate the mine, the existing pumps managed to keep the mine "in fork" until the new engine arrived finally and was erected during 1852. It was started, amidst great celebrations, on 16 September and the hero of the day was the Chief Engineer, John Congdon, who was rewarded with a gift of £50 for his work in the erection of the engine. Other, smaller items were also ordered from Cornwall. In May 1851, for example, SAMA placed an order for three copies of "... a work common in Cornwall, being Tables for ascertaining the quantity of water contained in ores ...," and objects as insignificant as crucibles had to be obtained from Bolitho's smelting works at Calenick, near Truro.

Just as the Burra had led the way in the production of ore and the acquisition of machinery, so it acted as a strong impetus to the development of smelting in the colony. In April 1848 the *Register* had written that "Smelt we must, here, ... (for) we are possessed of something more than a Cornwall ..." After a false start at the Burra in 1846, a successful smelting works was established at Port Adelaide by Dr. Edwin Davy (a relative of Sir Humphry Davy), and a short time later another works was set-up at Apoinga, just south of the Burra, with Captain John Rowe (a Cousin Jack) in charge. The Thomas brothers, from Hayle, arrived in the colony to erect smelters at the Callington and Kanmantoo mines, while small works were also built at Kapunda and Glen Osmond. A much larger works was erected at Burra Burra in 1848 and its owners, the English and Australian Copper Co., set-up in 1861 a second and even bigger complex at Port Adelaide. The growth in smelting was reflected, too, in the rise of engineering foundries in the colony - a number run by Cornishmen for the purpose of building machinery for the mining industry. The most famous of these was Martin's of Gawler. From small beginnings in 1848, James Martin (from Stithians) developed his foundry to the point where, at his death in 1899, it was employing 700 men and constructing some of the largest mine engines ever built in Australia. Other Cornish foundries included Hawke's of Kapunda, May's of Wallaroo, and Jones' of Adelaide.

The amazing success of the Burra Burra mine created an atmosphere of "coppermania" in the colony which prevailed from 1845 until Victorian gold stole the lime-light in the early 1850s. Despite the early failure of the Princess Royal, others were keen to try the country in the neighbourhood of the Burra. The Bon Accord was worked intermittently without success from 1846 until 1863, and the Murray Scrub, Emu Flats, and Karkulto mines - all in the same locality - all similarly failed. In the Adelaide Hills "coppermania" resulted in the opening of Wheal Acraman, Wheal Boone, and the Worthing Mining Co.s claims - the latter worked by Captain Alfred Phillips who had been especially brought out from Cornwall. Most startling of the Hills finds were those at Tungkillo (the Reedy Creek, Great Wheal Orford, and Wheal Rothschild mines - worked by Cousin Jacks like Henry Adams from Tuckingmill, James Spry from Launceston, and John and Edward Dunstan from Wendron), and in the tract of mineralized country running from Mount Barker through to Callington. The latter district, with mines such as Wheal Rose, Wheal Fanny and Wheal Fortune, with villages called Callington, Kelynack, and St. Ives, and miners like Absolom Tonkin from St. Blazey, William Penhall from St. Austell, and John Curnow from Halsetown, became for a time "... the Cornwall of Australia."

One of the Callington mines, the Kanmantoo, was run in its early days by Captain Richard Rodda. Rodda also managed the neighbouring Wheal Maria, but was most especially known for his work in the Barossa Valley. He came from Penrice, near St. Austell, and arrived in 1846 to work the mineral deposits on the lands of George Fife Angas. He became one of South Australia's most active mining men, founding the mining village of Penrice and opening-up the Greenock Creek, Lyndoch Valley, Wheal Friendship, and North Rhine copper mines. He was also associated with the Crinnis Mine, a working "... named after the celebrated Crinnis Mine of Cornwall ..." Wheal Barton was another copper mine worked on Angas' property, at the northern extremity of the Barossa district near Truro - a township "... named after the Cornish glory," to quote one contemporary newspaper report. The captain, however, was not Rodda, but John Rowe-who had relinquished his position at the Apoinga smelters in 1849 to go to the mine. A few miles down the Bremer River from the Callington district was Strathalbyn, yet another area with Cornish men and Cornish mines until the discovery of Victorian gold led to the disappearance of the former and the demise of the latter (in 1854).

The Victorian Gold Rush, indeed, brought a decade of extensive, almost feverish mineral development in South Australia to an end. By December 1851 the smaller mines were already closing. Kanmantoo was abandoned in 1852, as was Kapunda, and even the mighty Burra Burra ground to a halt. In March 1851 there had been 1,042 men working at the Burra mine. This figure declined to 360 by March 1852, to 157 by the following September, and was soon less than 100 - only a handful of these being underground miners. In November 1852 Henry Ayers wrote that "... the operations of this company are in great measure suspended from the want of the necessary labour ...," and ore raised fell from 23,000 tons in the year ended September 1851 to a low of 2,000 for the year ended September 1853.

The recovery of the South Australian mining industry did not ocurr until further supplies of miners arrived from Cornwall and, more particularly, until the men began to return from the Victorian gold-fields. Of the numerous mines that had been in operation in the Adelaide Hills before the rush, only Wheal Watkins was ever again worked successfully. But Cousin Jacks such as the Captains Paull, Pascoe, Rodda, Penberthy, and Barkla, tried their best to make going concerns again of mines like the Reedy Creek, Kanmantoo, Callington, and Wheal Friendship, while several newly discovered claims were started during the 1850s. The most important of these was Wheal Ellen, a silver-lead mine in the same district as the earlier Strathalbyn mines. The working was opened in 1857 and fully active by 1860, its owners determined to run it true Cornish style. To this end, they acquired a 60 inch Cornish engine and brought out several experienced Cornish mining men - John Cornish from Helston, Thomas Nicholls from St. Austell, Henry Richard Hancock from Horrabridge (in Devon), and William Arundel Paynter from Hicks Mill, Gwennap. Captain Paynter had worked in mines in Cornwall and Cumbria before migrating to South Australia, and it was while in Cumberland that he had first heard of Wheal Ellen. He wrote to his wife Sophia, mentioning that "... there is a cornishman here about to leave for Australia on Tuesday or Wednesday next, his name is Vial from Chacewater ...," and a few weeks later - in February 1859 - he told her that "... I have had a little information as to the mine in Australia ..." By the September he was at Wheal Ellen, saving hard to bring out Sophia, and working as chief ore-dresser and engineer. He stayed at the mine until 1861, then moving on to the newly discovered Yorke Peninsula mining district.

At Kapunda, which at the height of the Victorian rush could boast only four employees, work began again in 1854-55 as the miners returned. The mine then produced an average of 3,000 tons of ore per annum for the next 10 years but, by the early 1860s, falling copper prices and increasing pumping difficulties made it uneconomical. A change of ownership in 1865 made little difference, and from 1870-1878 only 1,250 tons were being produced annually - much of this through open-cut methods and the re-treatment of wastes - the mine then being abandoned forever.

The Burra fared only marginally better in the post-rush era. By 1865 it had also lost its strong economic base (as had the copper industry in Cornwall). A loss of £19,000 was suffered during 1865-66, a prelude to the disasterous years ahead when, between 1872 and 1877, SAMA lost £30,000 while producing only 14,000 tons of ore. This decline was particularly unfortunate because SAMA had responded to each period of crisis in the years 1858-78 with innovation, especially the introduction of new machinery. In 1857 it had ordered another 80 inch engine from the Perran Foundry, while in 1858 it sent Captain Bryant to prospect in the Far North of the colony and negotiated with James Trestrail, a Cornishman who claimed to have found copper on his "Trevale" property in the Adelaide Hills. In January 1860 Ayers instructed Thomas Williams of Moretonhampstead, Devon, to glean details of new methods employed at mines run by John Taylor and was most satisfied when he received "... sundry drawings of machinery in use in Cornwall, for hauling and washing ore ..." At the same time, Ayers initiated the re-treatment of the "slimes" that had collected in the creek, with the intention of boosting the overall copper output.

Despite this enterprise, the Burra was plunged into decline in 1862 through an unfortunate combination of slumping copper prices and an unusually wet winter which made it difficult to keep the mine dry. In

December 1864 Ayers described the state of the copper market, and its implications for the Burra, as "very deplorable," but during 1865 SAMA displayed a new-found enthusiasm - the first manifestation being the commissioning of a new survey of the mine by several of the colony's leading captains. This resulted in a wide variety of new exploratory work, and SAMA engaged Captain Isaac Killicoat to review completely the mine's ore-dressing processes. Killicoat was a man of some standing in the colony and, like Henry Roach, had once worked at the Tresavean mine in Cornwall - where he had been "grass captain." He had lived at Perranwell before migrating to South Australia in 1853 to take charge of the smelting works at the Burra.

By February 1867 the Burra mine was losing £100 a day, despite a series of drastic economies. In 1873 Captain Robert Sanders, another Cornishman, was appointed the new superintendent and given the unenviable task of trying to make the mine pay its way. He took immediate steps to recruit new miners from Cornwall, purchased new machinery, and instigated a thorough search of the deep workings - which revealed two new lodes. Development and exploratory work continued through 1876 and early 1877, but by the June a further fall in copper prices made necessary another series of drastic economies. Ayers warned the mine might have to be abandoned, a course of action that had become inevitable by the September. The machinery was shut-down and the employees dismissed on 30 November 1877. Sanders, despite his energy, optimism, and Cousin Jack flair, had failed to save the once-great Burra Burra. The *Burra Record* felt that the set-back was only temporary, but it was truly the end of an era.

The miners who had stayed to the bitter end moved on to other mining districts, but Burra did not lose entirely its Cornish connections for, in the thirty years or so preceding the closure, many Cousin Jacks had found alternative employment in the district's townships, while others had begun farming in the savannah-like hill country of the Lower and Mid-North region around Burra and Kapunda. In Kapunda township, as the *Royal South Australian Almanack and General Directory* for 1855 reveals, the Cornish were quick to involve themselves in a variety of activities - there was Philip Trenery, ostler; Edward Batten, shoemaker; John P. Moyle, horsebreaker; Nathaniel Hawke, butcher; William Pearce, innkeeper; Waters & Trevaskis, storekeepers; John Rowe, cattledealer; and so on. At the Burra the story was the same - Henry Nankervis, publican; William Oliver, stonemason; William Trembath, cobbler; Charles Rule, wheelright;...

The Victorian Gold Rush proved an important stimulus to wheat-growers in the Lower and Mid-North, thus encouraging many Cousin Jacks - especially those who had done well on the goldfields and now had the means to acquire land - to exchange the pick for the plough. The miner-turned-farmer became a familiar figure in the area, the change in occupation being not so radical as might appear for in the last century the division of labour was not so clear-cut, many Cornish miners at home having also been part-time farmers with rented small-holdings. In South Australia, their pioneering spirit could be quite staggering. John Curnow, for example, was a Kapunda miner before going to the Victorian diggings. After some success he returned to the Kapunda area and he had his sons purchased a plot of virgin bush land. Unable to afford expensive ploughing machinery, they set to with spades, digging four acres which later yielded them 160 bushels of wheat worth £1 per bushel. In the Burra district, James Sampson - from Goldsithney - was one of the first miners to turn to farming, while Henry Pinch arrived back from Victoria rich enough to buy his 784-acre "Pencarrow" farm.

At Spring Farm, to the north of the Burra, there developed an extensive community of former miners and their families. They became close-knit, inter-married, carved out large farms with names such as "Trevenson" and "Treview," and moulded their social life through the medium of the local Wesleyan chapel. Samuel Bray and his wife Martha Glasson were both from Falmouth. Thomas Moyses lived at "Treview." His wife Mary was the sister of William Blight (from Pool) who was in turn married to Jane Bone from St. Just. Jane's brother James also lived in the district, and her son - William - married Mary Dunstone, a daughter of John Dunstone of "Trevenson." Thomas Ninnes (from Towednack) and John Chapman were other Cornish settlers. John married Amelia Teague (from North Hill), while another of the Chapmans, William, married Thomas Ninnes' daughter - Martha Maria. As time went on, these relationships became all the more complex, Spring Farm earning something of a reputation in the colony as a Cornish-Methodist stronghold.

As the Lower and Mid-North became more and more densely settled by farmers, so those searching for pastures-new had to look even further north, to the marginal land along the southern Flinders' Ranges. In the Burra area, "Many of the Cornishmen, with little hope of expansion on small farms ... sold out and headed north," while the closure of the Kapunda and Burra mines in 1877-78 coincided with a new Government policy to push forward the frontiers of settlement and populate the Far North. Thus it was that Cousin Jacks, like Henry Bastian from Kapunda and Richard Ellery from the Burra, made their way to the little-known north, not a few succumbing to the dreadful drought conditions a decade later when many northern properties reverted to semi-desert.

However, despite all this, the main feature of the post-rush era was not the revival and demise of the Kapunda and Burra mines, nor even the consequent Cornish shift to the land, but rather the discovery and rapid development of the copper deposits on northern Yorke Peninsula. Here the growth of the mines was so indentified with the Cornish that the entire Moonta-Wallaroo-Kadina complex was soon known as "Australia's Little Cornwall," and again an overwhelmingly Cornish contribution can be traced. Copper had been noted on Yorke Peninsula by no less an authority than Captain Richard Rodda as early as 1848, but at that stage the region was too

41

remote and under-developed to warrant exploitation of its mineral deposits. Even in later years, when the district had become settled, the environment was in many ways unfriendly and uninviting still. In 1872 Captain Samuel Higgs of the Wallaroo Mines informed members of the Royal Geological Society of Cornwall that the Peninsula was a

... a tongue of land lying west of Port Adelaide, bounded on the east by Spencer' Gulf, and on the west by St. Vincent's Gulf. It is about 150 miles long, and varies in width from 25 to 50 miles. It is a country entirely destitute of fresh water ... the only natural vegetation is a stunted mallee tree-scrub, the foliage of which is a dirty olive green ... The grass, known as spear-grass, after rains, grows rapidly ... but a day or two of a South Australian sirocco (north wind) dries it entirely up, and the whole face of the country assumes the aspect of an arid wilderness.

It was in December 1859 that the first major discovery of copper on the Peninsula was made, at a place soon to be known as Wallaroo Mines. In 1861 the Moonta Mines, some 10 miles to the south-west of the Wallaroo find, were discovered, and it was not long before countless claims were being worked in the area. The Wandilta, the New Cornwall, the Doora, Wheal Burty, the Paramatta, Copper Valley, Wheal Devon, the Albion, the Poona, the Euko, Wheal James, and Wheal Hughes were only a few names amongst the 50-odd working mines recorded in May 1873, while the surnames Kitto, Bunney, Prideaux, Goldsworthy, Bice, Lean, Snell, Pascoe, Trenouth, Edwards, Gray, Northey, Vercoe, and Wearne, were just a sample from the long list of Cornish captains who managed the workings. Most of these smaller mines were soon abandoned, or absorbed into the larger Moonta and Wallaroo workings, but two which proved to be good "stayers" were the Hamley and the Yelta. The former began life in 1862 as the Karkarilla which was closed in the late 1860s but later re-opened as the Hamley and worked through to the First World War, when it was absorbed into the Moonta Mines. The Yelta, first discovered in July 1862 on the northern boundary of the Moonta leases, was worked until 1878 when it was abandoned due to the low price of copper. It was later worked sporadically, particularly in the 1890s, and was in 1903 acquired by a French company which was busy re-working a number of mines in the area.

Local place-names such as Tuckingmill, Helston, Jericho, Menadue, and Portreath, were a clue to Cornish involvement in the Peninsula mines, and indeed the very first miners in the district were Cousin Jacks from the Burra - William Pascoe, Richard Walter, Samuel Truran from St. Agnes, and Walter Phillips from Bokiddick, near Luxulyan. The expansion of the Moonta and Wallaroo ventures, not surprisingly, created considerable

42

excitement in the colony's other mining areas, and there was a steady movement from those districts to the Peninsula throughout the 1860s and 1870s. Thomas Cowling recalled that when he arrived at Yelta, near Moonta, in the late 1860s most of the inhabitants were:

... natives of Cornwall who had come to this country before the opening up of the Peninsula mines. Several of them came from the Burra, other from the Kanmantoo and Callington mines, and a few from Kapunda.

Others came from Victoria, with many more from Cornwall itself, and they fused with those arriving from the various South Australian mining districts to create an enormously vibrant and cohesive Cornish community on the Peninsula. The writer has been able to marshall biographical details of several hundred Moonta and Wallaroo miners, but it would be clearly impossible to note them all here. Suffice it to mention but a few. From the Burra came men like Edward Moyle (a native of Crowan) and Francis Manuel (a St. Blazey man) and John Tamblyn (from St. Agnes), while among the Kapunda Cornish were John Retallick (born in St. Austell in 1815) and John Isaacs (again from St. Austell). From Callington and Kanmantoo came the captains William Bray and Thomas Cornelius, and from Cornwall direct came a whole string of Peninsula-ites: Methusaleh Tregoning from Wendron, John Noble from Penryn, Charles Harris from Tywardreath, George Vercoe from Marazion, Joseph Bray from St. Austell...

Amongst those from Victoria deserving especial mention is Captain James Pryor. He was born at Rame in the parish of Wendron in 1845, and as a boy worked in the local mines - Retanna Hill and the Wheal Lovell group - before going to Victoria in 1866 to work the alluvial gold claims near Ballarat. He arrived at Moonta in 1869, and by 1873 was underground captain at the Poona mine. He later became a close collegue of the celebrated Henry Richard Hancock at Moonta and Wallaroo, and in 1908 was appointed Acting General Manager of the mines when H. Lipson Hancock was touring overseas. But perhaps his major claim to fame was that he was the father of Oswald Pryor, the Peninsula's cartoonist and author who wrote *Australia's Little Cornwall*. One miner who remembered James Pryor wrote that,

He was an austere, grey-bearded man, self-educated and self-contained. Lipson Hancock stood somewhat in awe of him; but he was his greatest support, strong in those points on which Hancock was weakest.

Many others could be singled-out for special attention, but of particular interest are the captains Thomas and Richard Cowling - one of the colony's celebrated father-and-son mining teams. Thomas was born in

Baldhu in West Cornwall in 1827, but later moved eastwards to Gunnislake to work at Wheal Edward. He spent some time in North America, but in 1862 emigrated to South Australia with his friend Captain East to run the New Cornwall mine. He later worked at the Wheal Hughes, Yelta, North Yelta, and Hamley mines, retiring from the latter in 1884. He was succeeded there by his son Richard, who had been second captain since 1878. Richard was born at Gunnislake and as a lad had worked in Drakewalls Mine before joining his father in South Australia in 1867. Like his father, he had worked variously at Wheal Hughes, the Yelta, and the North Yelta, but it was with the Hamley that his name was always connected.

By 1875 it was estimated that there were some 20,000 to 25,000 people on northern Yorke Peninsula, mostly Cornish immigrants and their descendents, and these settlers were responsible for the dramatic rise of the Moonta and Wallaroo mines. And, as at Burra and Kapunda, the Cornish soon settled into other occupations, becoming artizans and farmers in the localities. At Moonta we find William Chappell, cobbler; James Trezona, storekeeper; James Jewell, plumber; while at Wallaroo there was William Madden, mason; William Trelease, smelter; John Oates, teamster; and at Kadina John Liddicoat, carter; Anthony Sando, bootmaker; Samuel Mutton, teamster. Land in the district was first made available for farming purposes during 1866, and by July 1870 the Adelaide *Observer* could note that the locality

... is gradually being occupied for agricultural purposes. There are thousands of acres of good arable land which are yet destined to blush with the fresh verdure of cereal crops. Farmers are struggling manfully to supplant mallee scrub with cornfields.

In the early 1870s excellent harvests increased demands for more land to be made available, with the Peninsula miners in 1874 presenting a petition to the Adelaide Parliament. Several emphasised that they had been small-holders at home, one exclaiming that he "... had been a little bit of a farmer in Cornwall, which was the next paddock to England." Governments in the 1870s were responsive to such demands, and by 1878 the Peninsula's wheat output amounted to one million bushells, worth £250,000. Again, one could list scores of Cornish miners who turned their hands to farming on the Peninsula, but a few examples will have to serve as general illustrations. That the process of actually "winning the land" was difficult in the extreme is shown by contemporary accounts, one recalling that Benjamin Rose "with not a few others from Moonta Mines ... tackled the mallee country for which Yorke Peninsula is famous ... (a task) which might have been described as penal servitude for life." Similarly, Charles Wesley Bowden recalled that "My father, and others from Cornwall, had to do the job properly or not at all. But

I don't think they had any Black Mallee roots to contend with in the Old Country." Some indeed were not successful, like poor Elisha Mayne who - while trying to support two elderly parents back in Cornwall - was attempting to develop a farm south of Kadina. He struggled on from 1875 until January 1877 when, succumbing to drought and debt, he was forced to sell his land and abandon farming. Throughout the 1880s, a time of drought and economic depression, many other Cornishmen were also forced to leave the land, but those who survived were lucky - for by the early 1900s Yorke Peninsula had become established as a prime wheat-growing district.

However, despite the growth of farming, it was the Moonta and Wallaroo mines which provided the economic backbone of the Peninsula in the last century. Ultimately, through economic pressures, the two mines were forced into amalgamation, but until 1890 they were independent concerns, albeit operating in close co-ordination. In its 30 years of independence the Wallaroo Mines raised 491,000 tons of copper ore, averaging at 11 percent, the most productive years being from 1866 to 1875. Due to low copper prices, a loss of £26,000 per annum was sustained between 1876 and 1880, and, although there was a brief revival in the early 1880s, economic prospects remained gloomy. For the entire period 1860-1889 there was an overall profit of £91,000, but after 1876 the Wallaroo Mines were a marginal concern and they continued in production - if somewhat erratically in the '80s - as the profits from the smelting works at Wallaroo compensated for their losses, the Directors always hoping for a rise in copper prices to restore profitability.

In those early decades, the Moonta Mines were economically stronger than their Wallaroo counterparts. In the period 1862 to 1889 production averaged 19,000 tons per annum, with the ore at 19 percent copper, but costs rose and combined with low copper prices to produce the Moonta Mines' first losses in 1878-79. Profitability was regained in the early 1880s, but by the end of the decade the mines were again in financial trouble. The common economic problems of the Moonta and Wallaroo companies led to their amalgamation in 1889-90, it being reasoned that a large firm would be better able to withstand the pressures of adverse market conditions, while at the same time benefitting from economies of scale. This premise was shown to be correct, the amalgamation breathing new life into the mines which was to last until 1923. Together, the Wallaroo and Moonta mines - before and after their amalgamation - dominated the colony's mining scene, in 1873 Anthony Trollope noting that "... when men talk of the mining wealth of South Australia they allude to Wallaroo and Moonta."

This position of pre-eminence, and the ability to survive in lean years, derived in part from a willingness to mechanise and innovate. The Wallaroo Mines in July 1864, on the recommendation of Captain East (from Calstock), ordered a 22 inch rotary beam-engine and accessories from the Bedford Foundry, an engineering establishment at Tavistock, Devon, in what was in so many ways an eastward extension of the East Cornwall mining district. East was replaced by Captain Dunstan, whose preference was for machinery produced by William West & Co. of St. Blazey. On his advice a vast array of equipment was purchased from West's during 1866-67, while a decade later a 60 inch pump engine was purchased from Harvey & Co. of Hayle. All manner of other equipment, from pump leather to Cornish shovels, was also aquired direct from Cornwall.

The Moonta Mines, like its Wallaroo counterparts, began to organise the import of machinery and materials from Cornwall at an early stage. In February 1862 it ordered a 60 inch pump engine from Harvey & Co. and purchased other equipment from the Foundry, although later - following a wrangle concerning a further engine the company had ordered, then cancelled - items were obtained from the Bedford Foundry and, in 1880, from Messrs. Hathorne-Davey, a partly-Cornish owned firm in Leeds. Other necessities, from rope and chain to rivets and pressure-guages, were also imported straight from Cornwall, with costly theodolites coming from the firm of W. Wilton of St. Day.

If some of the credit for the development of Wallaroo and Moonta must go to the Cornish foundries which produced their machinery, then equal attention should be given to the fascinating managerial regime of Henry Richard Hancock. Hancock was born at Horrabridge, just over the border in Devon, in 1836. He worked as a youth in the Cornish-dominated Tamar Valley mines, and so the Cornishmen at Wallaroo and Moonta - while never forgetting that "... the manager of the Moonta is a Devon Dumpling..." - considered that though he was not a proper Cousin Jack, he was "... near enough to one." Hancock migrated to South Australia in 1859, to work at Wheal Ellen near Strathalbyn. In 1862 he moved on to Yorke Peninsula where he was employed on short-term contracts by several of the mines until, in June 1864, he was engaged as Chief Captain by the Moonta Directors on the fabulous salary of £350 per annum. He immediately applied himself with vigour, recruiting new staff, arranging for further supplies of miners from Cornwall, and formulating a plan for the rapid expansion of the mines. His bold policies brought him into conflict with the more cautious Directors who considered him reckless, but Hancock's enthusiasm and charisma always carried the day. He used his innovatory skill to develop equipment such as pneumatic drills, and gained immortality in the mining world through the patenting of his spectacularly successful "Hancock Jigger" - used in the processing of ores. By the early 1870s Hancock had more real power than the mines' Directors, on several occasions over-riding the Board's decisions and then successfully defending his actions.

In April 1877, on the resignation of Samuel Higgs, Hancock became Chief Captain of the Wallaroo Mines, whilst still retaining his post at Moonta, thus further increasing his power, prestige, and influence. He responded to the low copper prices of the late 1870s with a prudent plan of ore conservation and, again to the dismay of many of the Directors, by pressing ahead with new capital investment. By the early 1880s Hancock's mastery was complete, the Directors admitting openly their indebtedness to his courage and determination, and in 1889 he became General Superintendent of the amalgamated Wallaroo and Moonta Mines - a position he was to hold until 1898 when he was succeeded by his son, H. Lipson Hancock. He retired to his home near Adelaide, where he died in January 1919. His reputation, of course, was tremendous. But because his attitudes to life and work had been so very serious and rather strait-laced, many jokes and yarns had circulated in the Australian mining camps, asserting that in his private life he was riotous and - to quote Oswald Pryor - "...a dreadful libertine." A few people believed the stories, but anyone who had met Henry Richard Hancock knew they were untrue. One person who knew Hancock intimately in his later life wrote that he was:

... a benign, white bearded patriarch with an old-world courtesy, and I came to the conclusion that the numerous stories about him ... were myths. That he was a masterful man, there is no doubt.

In marked contrast to the power and success of Captain Hancock at Moonta, were the dismal careers of the Cornish captains at Wallaroo Mines in the years before 1877. Eneder Warmington resigned as Chief Captain in 1864 after a miners' strike, and, following the temporary regime of Captain East, was replaced by Captain Edward Dunstan who arrived from Cornwall to fill the post in January 1865. He spent much of his time, however, feuding with his second captain - Paul Roach - and, following repeated warnings from the Directors, was dismissed finally in early 1869 as a result of his persistent "... improper sampling of the Tributers' Ores..." He in turn was replaced by Samuel Higgs, an eminent Cornish captain and geologist (and son-in-law of Sir Humphry Davy, to boot), who arrived from Penzance during 1870. By 1872, in deference to his qualifications and standing, Higgs was on the amazing salary of £850 per annum. But, as the decade wore on, the Wallaroo Directors found disconcerting the fact that their mine was becoming increasingly unprofitable while the Moonta company was still making a healthy surplus. They felt that management must play some role in determining performance and began to wish that they had secured the services of Captain Hancock - who was by now making quite a name for himself. Accordingly, in April 1877 the Board sought the resignation of Captain Higgs, explaining that:

The reason for this decision is that the Mines are at present working at a heavy loss and rather than close them altogether at once, the Directors wish to try if under other management more successful results can be obtained.

Captain Hancock, with his emphasis on ore conservation, claimed that "Higgs ripped out everything..." of value, and that it would take twelve months of developmental work to put the mines on a secure footing. Higgs, then, despite his excellent Cornish credentials, was perhaps guilty of bad management and poor mining methods - "picking the eyes" out of the mine at a time of low prices when a policy of conservation would have been preferable. Elsewhere in South Australia it was possible to point to poor management and methods, and so - resisting the temptation to pour unqualified praise upon the Cornish miner - a more critical assessment of his skill is perhaps in order. Certainly, there were criticisms of Cornish miners and their worth in the colony. Their customs and superstitions were ridiculed, especially the belief that Jack o' Lanterns (faery lights) were supposed to illuminate the sites of new ore lodes at night. One newspaper article, appearing in the *Register* in 1864, took the form of a humourous report on a fictional "Wheal Bald Hill John Mining Company," asserting that the find was due to "... the miraculous interposition of Providence by a pious Cornishman, who walking up a Bald Hill by night saw a dog tail-piped with a lantern... a splendid surface indication..." A similar piece, again carrying an implied criticism of Cousin Jack, in the *Register* in 1872 was a sham prospectus for a "Great Distended Bubble Mining Company," its property said to be adjacent to "... the world renowned Wheal Barrow Mine...," the report furnished by a "Captain Trepolpen."

Some of the practices of the Cornish miner were based upon custom rather than reason (the site of the first shaft at Wallaroo Mines was selected by a miner whirling a pick around his head, letting go, and then digging at the spot where it landed), and from time to time articles appeared in the colonial press attacking his conservatism. Cousin Jack also took a dim view of book-learned mining experts, one responding to an erudite geological report on Wallaroo Mines with the quip "... this may sound very grand to men who never saw a mine. But to a man that has been bred in Cornwall it is all humbug." But, superstition and conservatism apart, the Cornish miner's competence was sometimes also questioned. At the Hamley mine Captain John Warren was charged with "... gross mismanagement and incompetence...," although later exhonerated, and an authoritative report blamed the premature abandonment of the New Cornwall on the fact there had been "... considerable underhand stoping... and the hanging wall had not been secured properly with timber...," with result that the workings collapsed. There was also a feeling in South Australia that German miners, from the Hartz mountains, were more methodical and progressive than their Cornish counterparts.

However, there were others who objected to "... this nonsensical twaddle against Cornish miners...," one Cornishman in the colony complaining to the *West Briton* in 1850 that unscrupulous individuals were passing themselves off as Cornish captains - even though they knew little of mining - and thus earning the Cousin Jacks whatever bad name they may have had in South Australia. A contributer to the *Wallaroo Times* in 1868 emphasised the point that when it came to the bogus captain "... Cornwall does not acknowledge him, but hands him back to South Australian speculators as theirs..." Indeed, many felt that poor management was usually the fault of the local capitalists (very few of whom were Cornish) who were more concerned to profit through share speculation than to engage in serious investment in the mines themselves. A surprisingly candid report on the Burra mine, compiled by SAMA in 1881 in the hope of selling the by then defunct workings, admitted that in the early days short-term profit maximisation was seen as preferable to long-term development with the result that the "eyes were picked out" of parts of the mine prematurely. J.B. Austin, a local mining expert, also blamed the colonial capitalists, writing in 1863 that "... the abandonment (of mines) has resulted from the want of sufficient capital... I am well aware there are many of our abandoned mines which would be considered very promising properties in Cornwall." A Wallaroo miner, writing to the *Yorke's Peninsula Advertiser* in 1877, felt that "... it is a pity that the petty capitalists of South Australia should have anything to do with mining, for they evidently manifest an utter ignorance of its first principles."

There were others who were prepared to speak-up for the Cornish, J.B. Austin commenting that "... though we are apt sometimes to laugh at Cousin Jack we might occasionally gain some useful lessons from him." And Hancock was not the only innovator in the colony - Captain Paull at Wheal Margaret, Captain Anthony at the Kurilla, and Captain Paynter at Moonta were all well known for their technical contributions to the development of ore-stamping and dressing machinery. In the same way, leading Cornishmen in the colony were associated with the foundation of the Adelaide School of Mines at the end of the century, and the mining companies themselves expressed practical approval for the Cornish by actively recruiting them from Cornwall for almost 50 years. Whatever their shortcomings, the Cornish were prefered above all other classes of miner in the colony.

Indeed, the development and survival of mines in arid, remote, outback country such as Eyre Peninsula and the Far-North was a tribute to the tenacity and flair of Cornish captains and miners. The discovery of the Yorke Peninsula mineral deposits in 1859-61 had co-incided with similar finds elsewhere in the colony, and ushered in a second era of "coppermania." Cornish captains like Thomas Rodda (who in Cornwall had worked at Wheal Elizabeth in St. Merryn and at the Old Moor Mine, St. Austell) fought their way through almost inpenetratable scrub to open mines such as the Broughton River and the Charlton. On Eyre Peninsula Captains Phillips, Prisk, Vercoe, and Barkla struggled to make going concerns of the Port Lincoln, Wheal Bessie, the Murninnie, and other mines. The most successful was Wheal Burrawing, which was worked intermittently from 1871 until 1877 by a succession of Cornishmen, the Captains Tonkin, Datson, Parkin, and

Penberthy.

In the Far North, a Mount Remarkable mine was worked as early as 1848, but the first major copper discovery was at Mochatoona in 1858-59. Captain John Rowe, formerly of Apoinga and Wheal Barton, went to open the mine with a party of Burra miners who hoped "... to find more remunerative employment there, though the Mochatoona is 300 miles beyond the Burra." During 1860 the Great Northern Mine was started, worked for a time by Captain Pearson Morrison - "... a gentleman of considerable experience both in Cornwall and America." Over 250 miles North of Adelaide was Blinman township (quite a Cornish community in its day), and in its vicinity were a number of mines owned by the extensive Yudnamutana Copper Co., the most important being Wheal Blinman (later known simply as the Blinman Mine).

Of all the minor mines in South Australia, the Blinman was perhaps the most successful. In its first period of operation, 1862-1874, it raised £250,000 worth of ore, supported a local population of some 1,500 people, and was the subject of inumerable optimistic reports in the London *Mining Journal*. During that period the Blinman was managed by the Captains Anthony, Pascoe, Terrell, and Paull, but despite their efforts the mine succumbed to its various difficulties - the lack of water (for both drinking and ore-dressing), the hardness of the rock, and the problems and cost of transport. In 1882, after the penetration of the district by the railway, the mine was revived - this time under the management of Captain William Treffry Bryant who was especially brought out from the Old Treburgett mine, St. Teath, North Cornwall, to supervise the re-working. He was proud to describe himself as "... a Cornishman... bred up a working miner ... a self-taught practical man...," and he employed like-minded Cousin Jacks - Edward Nicholls as pitman, S. Rogers as mason, R. Skewes as carpenter, and W. Curtis, J. Snell and others as tributers. Work, however, was erratic. The mine was developed vigourously through 1883 and 1884, but then abandoned until 1888. By 1889 it was again in production, this time under Captain Doble - also from Cornwall - the mine later being taken by a Tasmanian firm and worked into the early 1900s.

There were other mines of some note in the Far North. There was the Ediacara, which as late as 1890 was being worked by old Burra hands who, since the closure of their mine in 1877, had wandered the northern districts looking for employment. The Prince Alfred, worked for both copper and silver-lead, created considerable excitement in the early 1870s, and near the northern regional centre of Hawker were a number of claims - the most important being the Wirrawilka which was opened in April 1861 by Captain Lean, "... well known to all mining men in the colony as a competent and straight forward man." The Sliding Rock was the home of a number of Cornish miners in its hey-day, the early 1870s, and the Vocovocanna was in 1882 a mine which attracted the interest of at least one Cousin Jack who wrote that "The regularity of the side wall of this lode reminds me of the Cornish Copper Mines." Other noteworthy mines included the Leigh Creek, the Beltana, and the Mount Lyndhurst.

There were many itinerant Cornishmen who made their way from claim to claim in the Far North, men like Henry Paull who was described by his contempories as "... a well-known miner and prospector of the northern districts." But of particular interest are the wanderings of Captain T. Tregoweth. In 1874 he was manager of the small Pararra copper mine near Ardrossan on mid-Yorke Peninsula, and in August 1875 was noted amongst those attending the foundation ceremony of the Wesleyan chapel at nearby Maitland. By 1882, however, he was in the Far-North - as captain of the Wirtaweena mine. Before the year was out he had moved on to the Mount Rose mine, 56 miles NNE of Blinman, which he considered "... the richest copper mine yet discovered in the north." During 1882 and 1883 he pushed forward development work at Mount Rose, but by 1887 was working at the Mutooroo group of mines near the New South Wales border. He was involved with other local claims, such as Smith's Olary which he reported on in 1891, but he remained at Mutooroo until at least 1897 - his employees being Cornishmen with names like Poole, Roberts, Bray, Hancock, Menhenett, and Ellis.

Despite all this activity, by 1900 most of these remote mines had been abandoned and there were few workings of any consequence outside northern Yorke Peninsula. However, there remained a handful who continued to roam the outback areas - sometimes alone and occasionally in small groups, men like Henry Paull and Captain Tregoweth, prospecting in the old mining districts. Most had been in the colony for years, and some could remember the halcyon days of Burra and Kapunda. They were the "old-timers" of Australian mining lore, men who wandered across the bush, building their "humpies" and brewing their tea in billy-cans, men who often shunned the company of other human beings - delighting in their own solitude - but who could always spin a yarn or two about life in the mining camps back in the early days. They were a breed of men unknown in small, close-knit Cornwall, and yet - to many Australians - they were the archetypal Cousin Jacks.

CHAPTER 3

LITTLE CORNWALLS

Whether in Jo'burg or Butte City, Keweenaw or the Rand, the Cornish had a tendency to "stick together," welcoming each other as "cousins" and treating other Britons as "foreigners," as they did at home. This was no less true in South Australia where the Cornish had a reputation for being "clannish" and were recognised as "different" by other settlers in the colony. To some extent this was inevitable, for the mining towns attracted mainly Cornishmen and were for the most part remote, cut off from the main centres of population in South Australia. But for many, strong bonds had been forged even before landing in the colony as a result of living in close proximity to one another for several months at sea (Cornish families tended to be despatched in groups), while whole "extended family" relationships were recreated in South Australia as a result of extensive emigration from particular areas of Cornwall. And once in the colony, Cornish settlers were often at pains to seek out other Cornish folk, while others were brought together as a result of their common enthusiasm for establishing Methodist meeting-places.

Examples of all these phenomena can be traced. In the "John Brightman" in 1840 Philip Santo from Saltash and James Crabb Verco from Callington formed a personal friendship which in the colony blossomed forth into commercial partnership, political alliance, and joint involvement in the non-conformist Church of Christ. And for couples such as Henry Pinch from St. Mabyn and Frances Hicks from Mevagissey, or John Jenkin of Perranzabuloe and his young Redruth girl, courtships at sea resulted often in marriage at Port Adelaide. When Marmaduke Laurimer wrote home to Cornwall in the late 1830s, his letters were full of news of friends and relatives who had emigrated with him. William Prowse was excited when he encountered other Cornishmen - Thomas Matthews from Penzance, a Mr. Coles from Helston, and "... Geo. Archer that left Madron Churchtown." Similarly, James Sawle was pleased when he "... went into a quarry last week, there were two men at work from Perranwell ..." and was offered work "... with a master mason, who came out from the neighbourhood of Bodmin." In the same way, Joseph Orchard from Mawgan-in-Meneage was delighted to find lodgings in Adelaide with a Cornish Methodist, Mr. Thomas Barker: "He is from Wendron, a Class Leader." Samuel Bray, from Falmouth, explained that "We have met many who love God and his people also. We have a chapel as large as Budock Chapel... we have as large a chapel as that at Penryn nearly finished."

Indeed, the success and strength of religious non-conformity, particularly Methodism, in early South Australia was in no small measure a result of the extensive migration from Cornwall. In 1844 the Methodist

EMIGRATION

TO

South Australia.

Mr. I. Latimer,

(AGENT FOR SOUTH AUSTRALIA)

Having been requested to explain the principles of COLONIZATION adopted by the SOUTH AUSTRALIAN COMMISSIONERS with regard to this Colony, begs to announce that he will deliver

A FREE
LECTURE

ON TUESDAY EVENING NEXT, AUGUST 27, 1839,

At the King's Head Inn, Chacewater.

As the Lecture is particularly intended for the instruction and benefit of the WORKING CLASSES, it is hoped that all those who feel interested in the subject will give their attendance punctually.

The Lecture will commence at Seven o'clock *precisely*, and at the conclusion the Lecturer will be happy to answer any questions relative to the Colony. Mr. Latimer will be in attendance at the KING'S HEAD previously, to give information to any Laborer, Mechanic, or Artisan, who may be desirous of obtaining a FREE PASSAGE to the Colony.

Truro, August 19, 1839.

E. HEARD, PRINTER, &c., BOSCAWEN-STREET, TRURO.

"**Emigration to South Australia**", - a poster issued by Isaac Latimer in 1839. Courtesy Royal Institution of Cornwall.

FREE
EMIGRATION TO PORT ADELAIDE,
SOUTH AUSTRALIA

Married Agricultural Laborers, Shepherds, Blacksmiths, Wheelwrights, Sawyers, Tailors, Shoe-makers, Brick-makers, Builders, and *all* persons engaged in useful occupations may obtain a

FREE PASSAGE

to SOUTH AUSTRALIA, where they are within the regulations of the Colonial Commissioners.

A meeting will be held by Mr. LATIMER, for the purpose of seeing the applicants and deciding on their eligibility,

On *TUESDAY, October* 15, *at BODMIN; at Ten o'clock.*

In the meanwhile all particulars may be known on application to

Mr. L. LATIMER, Truro.

AGENT TO H. M's. COMMISSIONERS.

All letters must be *post-paid* or they will not be answered.

E. HEARD, PRINTER, BOOKBINDER, BOSCAWEN-STREET, TRURO.

"Free Emigration to Port Adelaide", - another of Latimer's early posters.

54

The HOOGHLY at Port Adelaide, June 17 1839. Artist unknown. Courtesy S.A. State Library

Philip Santo, from Saltash, a prominent early colonist   Courtesy S.A. State Library

James Crabb Verco, from Callington, with family - circa 1864.

Courtesy S.A. State Library

57

Kapunda Mine, looking west, circa 1875. Kapunda township is in the right background.

A section of the Kapunda Mine, circa 1875, showing mine buildings and miners' cottages

A panoramic view of the Burra Burra Mine in 1874

Courtesy of S.A. State Library

59

Wheal Blinman - the Blinman Mine - in 1876

Blinman township, situated in the heart of South Australia's arid Far North, in 1876

Courtesy S.A. State Library

Elder's engine-house, Wallaroo Mines, circa 1887. The 80 inch beam-engine was built originally by Harvey & Co. of Hayle for the New Cornwall Mine on Yorke Peninsula.

Courtesy S.A. State Library

62

An early mine on Yorke Peninsula, worked true Cornish-style with horse-whims and whips

Courtesy Miss P. Minhard

Cornish miners at Hughes' Shaft, Moonta Mines, 1894.

Courtesy S.A. State Library

Unloading ore into an underground storage bin, Young's Shaft, Wallaroo Mines, early 1900s.

65

Miners at a shaft plat, Wallaroo Mines, early 1900s, waiting for stores and tools to be sent from surface. Courtesy S.A. State Library

A double-decked man-skip at a shaft plat, Wallaroo Mines, circa 1914-18.

Methodist Chapel at Cross Roads, near Moonta, late nineteenth-century

Courtesy S.A. State Library

population of the colony was 1,666. This figure increased to 22,210 in 1861, and by 1881 had reached an impressive 63,239. However, success was achieved not only through numerical strength, but as a result of Methodism's ability to adapt to - indeed exploit - the conditions associated with the colonising process. Unlike most other religious bodies, the Methodists did not need fixed buildings or paid clergy to survive, the local (or lay) preachers providing the impetus for Methodist expansion. Where no chapel existed they would hold services and class meetings in their own homes, and they were prepared to ride many miles through wild country to preach to settlers in outlying areas. A great many of these stalwart local preachers were Cornishmen, and as a new mineral discoveries led to the opening-up of hitherto uninhabited outback areas, so they were able to expand the Methodist sphere of influence. At local levels, Methodist organisation allowed for considerable freedom of action and this, combining with the Methodist committment to self-help, generated such activities as bazaars and tea-treats which enabled local societies -through their own initiative - to finance the construction of their own chapels. This was a situation reminiscent of that in Cornwall; and the austere, wayside chapel became as much a feature of rural South Australia in the nineteenth-century as it was of Cornwall. Of the 1,170 places of worship in the colony in 1900, no less than 608 belonged to the Methodists.

Another aspect of Cornish cultural influence which manifested itself quite early in the life of the colony was Cornish Wrestling. In February 1848, for example, the Adelaide *Register* carried the following advertisment:

I, William Hodge, weighing 12 stone 12 lbs, challenge any man in South Australia, to wrestle, Cornish or Devon style, two lock falls out of three to decide the challenge, for the sum of £20 or £50, the challenge to stand good for one month from this date. For further particulars, apply to Mr. Oatey, "Boars Head," Rundle Street.

William Hodge was the Cornish champion; another advertisement in November 1851 announcing a grand "CORNWALL V. DEVONSHIRE" contest to be held in Adelaide (there was to be seating for 2,000), with Hodge representing Cornwall and John Hoskin wrestling for Devon. Gawler, in those days the colony's second or third most important settlement, also had its wrestling matches - J. Wills being the Cornish champion in 1859.

The distinct Cornish identity remained strong through the years, resulting in February 1890 in the foundation of the Cornish Association of South Australia. Unlike the Cornwall and Devon Society of the 1850s, which was a pressure group designed to win specific advantages for the Cornish (and Devon) community, the Cornish Association was fundamentally a socio-cultural body, with strong nostalgic and romantic overtones. Its two leading lights were James Penn Boucaut and Sir John Langdon Bonython. Boucaut, born in Mylor in 1831, was a Supreme Court Judge and former Premier of the colony, a proud "... son of Cornwall ... (as) he has more than once boasted... on public occasions." Bonython was a newspaper magnate (owner of the Adelaide *Advertiser)*, amateur Cornish historian, public benefactor and philanthropist, and liberal politician. He had two fine houses built in Adelaide, Carminow and Carclew, named after ancient aristocratic seats in Cornwall with which the Bonythons of old had been associated, and used every opportunity to vent his passion for all things Cornish. The *Cornish Magazine* said of Boucaut and Bonython that "Both are enthusiastic Cornishmen," while news of the foundation of the Cornish Association was widely reported in Cornwall, accounts finding their way into such obscure journals as the Penryn *Commerical Shipping & General Advertiser For West Cornwall.*

The Cornish Association, however, was a somewhat self-conscious expression of Cornishness, and to observe authentic manifestations of the Cornish cultural inheritance one had to visit the colony's mining towns where the true Cornish identity was displayed in the everyday life of the ordinary people. Northern Yorke Peninsula was known as "Australia's Little Cornwall," but Burra Burra, Kapunda, and other smaller mining settlements were all "Little Cornwalls" for a time - relatively remote but vibrant and cohesive. It was these towns which gave the colony its far-famed Cornish heritage and achieved for their inhabitants their "clannish" reputation.

Kapunda was the first of these "Little Cornwalls." The miners kept their own feast days, such as the Duke of Cornwall's birthday, and formed their inevitable brass bands. Their wives baked the traditional Cornish fare, one old Kapunda resident recalling in 1929 that his

... earliest recollections of Kapunda go back three quarters of a century... a plate of Saffron Cake from the oven, and the vivid yellow colour of which I can still often, through the lapse of so long a time, plainly visualise.

Kapunda's *Northern Star,* first published in 1860, carried "News from Cornwall" items, along with amusing dialect stories written by locals with an obvious knowledge of both Kapunda and Cornwall. Even in 1879, after the closure of the mine, Kapunda was in the forefront of a relief campaign mounted in South Australia on behalf of the destitute of Cornwall. Locals felt that "South Australians should support the Cornish miners in their distress, for to the energy of some of them... the colony owned its prosperity," and they contributed £300 towards the relief rund.

But the Cornish did not have it all their own way, for there was also a strong Irish element in the town. Indeed, one of the features of Kapunda life was a continuing rivalry between the Cornish and the Irish. This may be interpreted as a clash of Celtic temperaments, although- with the Cornish

70

being the "labour aristocracy" at the mine, the captains, tributers, tutworkmen, and the Irish having the more menial labouring tasks - there was an element of class antagonism. There was also a religious aspect: the Cornish were often Methodists and sometimes Orangemen, and thus fiercely opposed the Catholicism of the Irish. On one occasion in 1862, when a group of miners complained that their ore was being unfairly assayed, threatening letters were sent to the chief captain and a public meeting held to discuss their implications. The Cornish implied that the Irish were responsible for the intimidation and, with the situation becoming perhaps a little ugly, several speakers found it politic to express "... their opinion that the writer was neither a Cornishman or an Irishman..." in an effort to diffuse the confrontation. Violence actually did break out some years later when, in the election of 1893, Patrick McMahon Glynn - the hero of the Irish community - was surprisingly defeated. Glynn blamed his defeat on religious bigotry, and one eye-witness recalled that "... the bone of contention was the Irish vote." Rioting followed, and the two men most badly beaten both bore Cornish names - James Rowe and W. Pengelly. When Michael Davitt, the Irish Land-League agitator, visited Kapunda in the 1890s he was surprised by the level of Irish sentiment, writing that it felt as though "... Kapunda was somewhere in Connaught..."

Thus although Kapunda was a "Little Cornwall," especially in the early days, the Cornish did not have a monopoly of the town. The various Burra townships (Kooringa, Redruth, Copperhouse, Aberdeen, and several smaller suburbs) also had a cosmopolitan strand, but the Cornish were numerically and culturally the dominant group. There were the usual Cornish-Irish clashes - political in the elections of 1851, but turning to violence in August 1867 when there was a fight between rival gangs of workers "... in which the Cornishmen got the worst." The Cornish were well aware of their seperate identity, the Rev. Charles Colwell (a Cornishman) telling Wesleyan chapel-goers at Kooringa in March 1859 that they were "... the real descendents of the Celts ..." Those brought-up at the Burra took a pride in their inherited sense of "difference," one local poet writing:

'Twas Grannie who lived near the old Burra Mine
And we often went up on a Sunday to dine
    At Redruth, where Grannie lived. Grannie would tell
    Us stories of Cornwall when she was a "gel,"
Oh, not in the world was a country so fine
    As Cornwall.

This same poet, "Cousin Sylvia" she called herself, in another of her works - "Lucky Find, 1845" - drew a link between the mystical qualities of Cornwall's Madron holy well and the almost miraculous discovery of the Burra Burra mine, concluding with the triumphant stanza:

And, oh, ye emerald Malachites!
Ye azure deeps of Madron!
A harvest of five million pounds
Was taken from those Burra mounds -
What luck! What chance! What fortune!

Burra did not get its own newspaper until 1876, the year before the mine closed, but when the *Northern Mail* (soon renamed the *Burra Record)* first appeared it carried numerous Cornish items. The Burra, too, responded well to pleas for the relief of the destitute in Cornwall, and local customs reflected the Cornish inheritance. The Duke of Cornwall's birthday, as at Kapunda, was observed as a general holiday, and there were the traditional miners' bands. Of particular significance was the popularity and survival of Cornish Wrestling. One report in 1848 described how up to a thousand people would gather to watch the events, while another in 1859 recounted in some detail the progress of an extensive series of matches held over three days "... in the real Cornish style." One commentator wrote disapprovingly in 1863 that the miners had spent Christmas "... lounging round the taverns, playing skittles and wrestling...," and there is evidence of the survival of Cornish Wrestling at the Burra until at least 1874.

There is also some evidence to suggest the celebration of St. Piran's Day (March 5th) in the early years at Burra Burra. And the erection of "Johnny Green," the miners' mascot - a life-size figure cut from iron - on the shears above Morphett's Shaft was in all probability the echoing of an earlier Cornish custom when branches and bushes ("St. John's Greenery") were set atop the shears on St. John's Eve. The feast of St. John on June 24th was the Christian form of the ancient Celtic ritual of Midsummer's Eve, still celebrated in nineteenth-century Cornwall by the lighting of countless bonfires. This was yet another tradition brought to South Australia by the Cornish miners, and which survived at the Burra for many years. Although falling in Midwinter, June 24th was kept as the "... red-letter day in Cousin John's calender..." and was, according to one report in 1863, "... celebrated by divers juveniles who lighted up numerous bonefires (sic)." Groups of miners also let-off "... logs of wood and hundredweights plugged with blasting-powder ..." - an obvious survival of the practice in Cornwall of "shooting" (blasting) holes in rocks to celebrate Midsummer.

The miners at Callington and Kanmantoo also kept Midsummer's Eve, and in 1859 the Callington mine owners chose June 24th as the occasion for christening their new 60 inch Cornish engine, an event they celebrated with a great dinner in the nearby "Tavistock Hotel." Similar festivities attended the starting of the massive 80 inch engine at the Burra in 1852, and also in June 1850 at Tungkillo when - according to the diary of one Reedy Creek miner - "Tables were arranged in the timber yard at the mine for 120 persons, who partook a bullock, roasted whole..." These "Public Dinners," when mine workers sat down to feast side-by-side with the owners and local dignatories, remained a feature of South Australian mining tradition until at least 1882, when an enormous dinner was held to mark the re-opening of the Blinman mine, and were derived from the very similar "Count House" dinners in Cornwall.

In contrast to both Kapunda and Burra Burra, the "Little Cornwall" community of Moonta, Wallaroo and Kadina on northern Yorke Peninsula was more homogeneously Cornish, and it is there that the creation and development of cultural patterns can be traced most successfully. It was many years before a migrant to Australia from Cornwall could feel completely at home. Some were home-sick, and others like Thomas Medlyn from Helston - who owned the smallholding of Polpenwith in Constantine parish - worried constantly about the state of whatever affairs, let alone relatives, they had left at home. The first settlers on Yorke Peninsula found it distinctly uninviting, Captain Dunstan writing that his wife refused to join him at Wallaroo Mines - "... a whim rope wouldn't be strong enough to draw her here." But in spite of all this, the district was transformed rapidly into a "Little Cornwall," newcomers then quickly adjusting to the scene and becoming part of the local community. One "new chum" recalled the uncertainty he felt as he journeyed to the Peninsula. But his fears were dispelled when the coach drew up in Kadina, for

... someone called out, who's that - old Bill? - how is Redruth looking, meaning the... place I came from in Cornwall, and then another called out is there anyone from Camborne? Thus was my reception at Kadina. As I soon found out, I was not the only one from Cornwall, and I replied asking how is the bal looking... and the answer was, plenty of ore.

The three main townships, of course, counted amongst their populations a fair number of tradesmen of other than Cornish nationality, and at Wallaroo there were employed in the smelters a considerable body of Welshmen who had been recruited in the mid-1860s from Cwm Avon, Taibach, White Rock, and the famous Hafod Works. But the inhabitants of the mineral lease settlements, Moonta Mines, Wallaroo Mines, and their various smaller off-shoots, were overwhelmingly Cornish. To the visitor from outside, the area's Cornish identity was unmistakeable. In 1873 Anthony Trollope wrote that "... so many of the miners were Cornishmen as to give Moonta and Wallaroo the air of Cornish towns," while by 1876 the district had been dubbed "an Australian Cornwall." Another visitor, this time in 1889, wrote that the locals "... lived isolated from the rest of the colony, remaining more Cornish than Cornwall itself." Certainly, this geographic isolation protected the Cornish from assimilatory agents at work elsewhere in the colony, while at the same time fostering both local patriotism and a sense of local identity. Other South Australians regarded the "Peninsularites" as different. Crowds of curious spectators witnessed the arrival at Adelaide of Kalgoorlie-bound Cornishmen from Moonta during the 1890s rush, one young gril remarking to her mamma "Are those people really Cousin Jacks? I thought they had long tails." Similarly, at the end of the century a newcomer from Adelaide recalled bitterly that: "My first impressions of Moonta Mines was - what had I let myself in for? It was soon made clear that I was a foreigner with habits and opinions to be viewed with suspicion."

The Cornish themselves made frequent use of the phrase "Cornwall, near England," and to some degree remained intentionally distinct from others in the colony. The Hummocks range of hills served to cut the Cornwall-like shape of Yorke Peninsula off from the rest of South Australia and, like the River Tamar at home, was a physical and psychological barrier. "I'll send 'ee over the Hummocks" was a threat used to discourage many an unwelcome visitor. As if to reinforce the point, a local Bible Christian, the Rev. John Thorne, remarked in 1874 that,

He felt very much at home on the Peninsula, it was more like Cornwall, almost surrounded by the sea and insulated in position, the miners preserved the same rugged characteristics that marked them in Cornwall, and preserved their independence.

While it seems unlikely that any of the residents of northern Yorke Peninsula could have known anything more than a smattering of phrases from the old Cornish language, the Cornish dialect of English (which included many Celtic words) did persist - as did the Cornish accent, Cornish idiomatic expressions, and other linguistic peculiarities. Far too many Cornish words in regular use on the Peninsula in the last century have been identified to be recorded here, but they range from words such as "nuddick" for the nape of the neck, and "clunk" for to swallow, to "zawn" for a chasm, and "kroggen" for a sea-shell. Cornish phrases such as "screechin like Tregeagle" and "scrowled as a pilcher" were also in regular use, as were mining proverbs like "mundic rides a good horse." Naughty children were threatened that they would be "... took to Bodmin" (ie. gaol), and others knew the mariners' prayer: "God keeps us from rocks and shelving sands, and save us from Breage and Germoe men's hands." Another saying, reflecting the traditional enmity between those two Cornish villages, was also well-known on the Peninsula:

Germoe, little Germoe, lies under a hill.
When I'm in Germoe I count myself well;
True love's in Germoe, in Breage I've got none,
When I'm in Germoe I count myself at home.

74

An equally interesting survival, again with a distinctly Cornish flavour, was the rhyme which went:

> There were three sailors of Mevagissey
> Who took a boat and went to sea
> But first with pork and leeky pasties
> And a barrel of pilchards they loaded she.

Locally composed "Cornish poems" also appeared in the Peninsula press from time to time, as did numerous comical "dialect stories," and a popular song was the stirring ballad 'Trelawny" - the Cornish "national anthem" - of which paraphrashed versions on local current affairs were often published.

These linguistic and literary survivals were matched by the survival of various Cornish superstitions, prejudices and beliefs. Many believed in herbal remedies and "... would sooner be attended by an old dear Cornish women than by doctors." New-born babes on the Peninsula had their gums rubbed with brandy - apparently a legacy of Cornwall's former smuggling days - to ensure they would never die by hanging. "He may be drownded but he waant be 'anged" was the old mid-wives' saying. The Peninsula folk also believed in the elf-like "Buccahs;" and "Jack o' Lanterns" moved across the fields at night, indicating the position of ore bodies. On the strength of one such sighting near Moonta in 1863 a company was actually formed to exploit the ground which had been "illuminated." Some people trusted in the art of ore-divining, and as late as 1921 Captain W.H. Hayes of Wallaroo Mines discussed seriously the supposed powers of local diviners in a strictly objective, technical report he had compiled on the mines.

Peninsula children played Cornish games such as "Annie Mack," and both youngsters and adults delighted in the singing of hymns and Cornish carols. They loved the old Methodist tunes such as "Tell Me The Old, Old Story" and, despite the Temperance movement, a visitor to the district in 1873 could note that "... ever and again you will hear sounds of singing from convivial Cornishmen who congregate together over a social glass..." There were numerous local choirs (many of them male voice), together with a number of surprisingly talented local muscians and composers - the most famous being James "Fiddler Jim" Richards, from Perranporth. His major works, along with those of other local writers, appeared in a volume *The Christmas Welcome: A Choice Collection of Cornish Carols,* published at Moonta in 1893.

Brass bands were also popular. "Bargwanna's Band," the first formed at Moonta, was in constant demand to lead the parades of the local chapels, trade unions, and temperance bodies. And in later years several bands, especially the "Wallaroo Town Band," formed in 1895, received widespread acclaim throughout the colony. Again associated with chapel life, were the curious but popular "Sacred Dramas" - a sort of latter day survival of the medieval Cornish miracle play, but now performed within the Methodist chapel rather than the Plen-an-gwarry. To the non-Cornishman, these plays seemed gauche and naive, one critic of a performance at Wallaroo Mines writing that it was fortunate that renderings of "Moses" and "David and Goliath" were now to be found only "... in some remote villages in England, and then almost confined to Cornwall, near England."

The strength of Methodism and the temperance movement did not prevent the brewing of an alchoholic herby-beer called "swanky," nor did it stop the emergence of illegal "kiddleywinks" or "sly-grog shops." Pasties, of course, were a Peninsula speciality, and other foods such as "fuggan" (heavy cake) and jam and cream "splits" were also prepared by Peninsula housewives.

Local rivalries emerged, as they did in Cornwall, one newspaper correspondent complaining in 1875 that it was always "... Wallaroo and Kadina versus Moonta..." People from East Moonta were known as "Copper Tails" while those from Moonta township were "Silver Tails" (reflecting the supposed social superiority of the latter over the former), and fighting between rival gangs of youths - even when institutionalised into local sporting fixtures - could be quite nasty.

In 1865 the *Wallaroo Times* commented that "Of all people in the world Cousin Jack loves his holiday..." How right it was! Survey-day, when tributers and tutworkmen negotiated their contracts, was an excuse for general merry-making, as were both the Duke of Cornwall's birthday and Whit-Monday (a traditional miners' holiday in Cornwall). Midsummer's Eve was an especially important celebration on the Peninsula, with scores of bonfires and an "... amount of powder, dynamite and other explosives used by youngsters... almost enough to bombard a town." The starting of a new engine was often marked by a holiday for a mine's employees, and with the usual public dinner. At the starting of the New Cornwall 80 inch engine in 1866 Captain East invited all the local captains to the festivities, for "It was a good old Cornish custom to meet together on occasions like the present, and show a friendly feeling, although engaged on different mines." And at the christening of the Paramatta engine in 1896 the Directors "... provided... a plentiful dinner... for every miner or other person in connection with the mine."

As at the Burra, Cornish Wresling was tremendously popular for many years. The matches, held principally during the festive seasons of Easter and Christmas at the larger hotels in Moonta, Wallaroo and Kadina, were well-organised affairs - with often 50 or more people participating in the final play-offs. Needless to say, local rivalries were played out in these matches: On one occasion there were great celebrations when the Moonta champion, John Doney, defeated his Kadina rival, William Mitchell, "... in true Cornish style..." and in 1868 the legendary John H. "Dancing" Bray of Moonta defeated the Ballarat champion with a superbly executed "flying mare" throw

- "Then followed a shout such as might have been heard when Sebastopol was captured. It was said that 20 captains who were there declared they had not seen anything equal to it in Cornwall." As late as January 1883 there were still sixty practicing wrestlers living in the Moonta area alone, and even in 1895 it was noted that Moonta miners on the Western Australian goldfields were participating in wrestling bouts.

But as time went on customs such as wrestling gradually fell into disuse. The passage of time was the main catalyst in cultural change, but children born on the Peninsula were also subjected to deliberate homogenising and assimilatory processes. The introduction of compulsory education in 1875 helped to erode the Cornish outlook and mannerisms of the local children, and some schoolmasters warned that their pupils "... must not make use of any Cornish expressions..." Some Peninsula residents resented this dimunition of the local identity, one writing angrily in 1883 that "So essentially Cornish are we that we would think not only our habits and national amusements would prevail, but that our Mayor, magistrates, and leading men would be chosen from among them..."

Cultural change, however, was not merely a matter of the Peninsula's Cornish heritage being continually eroded. There was, in addition, the growth of an increasingly romantic-nostalgic view of Cornwall (so different from the grim reality which had earlier led to the widespread emigration), together with the emergence of a new outlook in which the Peninsula itself became the primary focus of Cousin Jack loyalty. The formation of the Moonta and Kadina branches of the Cornish Association of South Australia in 1890 was the most obvious example of romantic-nostalgic change, but more than a little locally-generated local patriotism showed through in one Moontaite's comment that "We should think Moonta not only a good place to establish a branch, but that it could not be bettered..." Indeed, the whole district combined to agree that "If you haven't been to Moonta, you haven't travelled," a phrase Oswald Pryor attributed to the boastings of South Australia's Cousin Jacks at Broken Hill in the 1880s, but which in fact had been heard even earlier in the North American mining camps.

For many people, Moonta was the focal point of Cornish influence in Australia, its very name almost synonomous with "Cornishness," and people born on northern Yorke Peninsula developed an unusually strong emotional attachment to their home district. One of South Australia's minor poets, Thomas Burtt, offered an intensely personal interpretation of his own attachment to Moonta in a poem, "The Solemn Moonta Mines," which he wrote on return from several years of absence, the ... dirge-like sound and muffled reverberations ..." of the mines at night awakening in him a sensitive appreciation of the unique, rather awesome atmosphere of the place:

> Hark! methinks I hear the echo!
> Of those solemn Moonta Mines;

Sadly sounding distant far-off,
  Over flow'rets, trees and vines.

Listen to the ceaseless throbbing,
  Of those engines measured slow;
Telling many a weary spirit
  How it shares a world of woe.

Burtt was fortunate in that he had a degree of literary talent, so that he could express his innermost feelings in verse. But the average Cousin Jack, poorly educated and not given to poetic outpourings, had to articulate his sentiments of Moonta loyalty through the insistent repetition of his catch-cry "If you haven't been to Moonta, you haven't travelled."

The Cornish character of all these "Little Cornwalls" derived not only from the survival and development of cultural traditions from Cornwall, but also, as already intimated, from the social and religious conditions established in the mining towns. To an extent, of course, all three were related - and they all bore close comparison with conditions that obtained in Cornwall. In certain respects, social conditions in the South Australian mining towns were notably superior to those in their Cornish counter-parts. As noted in Chapter 1, wages were considerably higher for Cornish miners in the colony than could be had at home - a view confirmed by the writings of one Johnson Frederick Haywood, who in 1847 said of the Burra:

Of Beef and Mutton, the primest joints sold at ld per lb, and wages were so high at tutwork or tribute that a Miner rarely came away from the Butcher's Store with less than 20lbs of meat on his shoulder.

This contrasted strongly with the situation in Cornwall, where as late as 1864 it was claimed that miners could not afford more than $3\frac{1}{2}$ lbs of meat a week.

But despite the high level of wages and the availability of good food, other aspects of local conditions could be as bad as those at home. At the Burra, for example, the settlers had to contend with poor housing conditions, the prevalence of disease, and what might generally be termed the rigours of colonial life. Although the South Australian Mining Association (SAMA) was able to build some cottages for its employees, many Burra miners adopted the extraordinary practice of hollowing "dug-out" homes into the steep banks of the Burra Creek. One report noted that the banks "... swarmed with people like a rabbit warren..." from Redruth to Kooringa, and the *South Australian News* wrote in April 1848 that many "dug-outs" were "... fitted up in the neatest style imaginable and form neat and comfortable habitations..." They did, however, have their disadvantages. Practical jokers walking along the tops of the banks could drop objects down the make-shift chimneys, or even fish up pots boiling on the stoves! A more serious problem was, as one visitor noted in 1851 (when the population of the "dug-outs" was 2,600), that "... infantile diseases are greatly prevalent." Also, the creek was regularly inundated with flood-water, and after a particularly bad flood in June 1851 - when one miner lost his life in a collapsed "dug-out" - these subterranean dwellings were largely abandoned.

Disease, however, remained a constant evil. In 1847 the "bal surgeon" complained about the unhygenic conditions of local slaughter-houses and about the fact that "chamber filth" thrown from the cottages deposited in the creek. There were periodic outbreaks of typhoid, and in 1872 Kooringa was said to be "... the shabbiest town in the Australias ...," its cottages "... the most squalid in the British Empire ..." The behaviour of the settlers themselves, especially in the early days, was often characterised by violence and degredation, which had a generally depressing effect on social conditions. In October 1846 one report alleged that at the Burra "... drunkeness exists ... to a frightful degree. Wages are so high that the men have the means of gratifying their worst passions and it would seem ... that they are doing so." Frederick Haywood was of the same opinion, and recorded in his diary in 1847 that at one of the local inns (there were 13 all told) the customers amused themselves by breaking windows and fighting, with the landlord having to clear the bar at night with a cricket bat.

In August 1849 another report complained that "... the brutalising amusements called prize fights are rife at Kooringa and its neighbourhood..." By that time, though, local inhabitants were themselves taking steps to improve conditions in the district. During 1849 a party of Burra miners submitted a petition to the Colonial Secretary in Adelaide, drawing his attention to the immorality and lawlessness that existed at the Burra and demanding that some action be taken. Foremost amongst the signatories were Thomas Ninnes (from Towednack) and Samuel Bray (from Falmouth), both of whom were Methodist pioneers in the area. Indeed, the local Methodists played a significant role in transforming the Burra from a raw and riotous mining camp into a peaceable, Christian, and broadly law-abiding community. Although compressed into only a few years, this was an experience similar to the transformation of Cornwall by the Methodists in the last century, and the reputation of the Burra miner - like that of his counterpart in Cornwall - changed from noteriety for violence, vulgarity and intemperence to renown for his civility, responsibility, and sobriety. Even during 1849 one observer claimed that there was "... a visible improvement in the state of society at the Burra ..." which contrasted with the situation a year or two before when the district was "... a hell upon earth." By the end of the year another report could make the amazing claim that "The almost total absence of crime among a population of nearly 5,000 inhabitants, speaks volumes for the morality of the people..."

Who, then, were these Methodists who could ring such startling changes? Principally, at first, the Bible Christians - a Methodist sect which had its origins in Cornwall and which had a ready appeal for the Cornish miner in Australia. In the early days there were no ordained Bible Christian ministers in the colony, and thus the organisation of the denomination in South Australia was in the hands of a few determined and dedicated local preachers. The most important of these was James Blatchford, a Cornish miner born at Stoke Climsland in 1808. An orphaned child, he went to live in a public house where "Dancing and wrestling were favourite pastimes." But at the age of 26, on the death of his first wife, he was saved from this life of fruitless frivolity and converted to the Methodists, not long after marrying a Bible Christian girl called Charity Jury. Together, they emigrated to South Australia in 1847 where James went to work in the Burra mine. Soon they were joined by other like-minded Bible Christians - Thomas and John Pellew, John Halse, Thomas and Mary Richards from Marazion, and John Vivian from St. Austell - and before long they themselves had built a chapel on land granted to them by SAMA.

The Bible Christian Missionary Society was much impressed by this progress, and responded to South Australian calls for trained guidance by sending two volunteers to the colony in 1850. One was James Way, a Devonshire man, but the other was the Rev. James Rowe from Penzance. Once in Australia, he made his way to the Burra - rejoicing there in "The loving welcome and subsequent hearty co-operation of the splendid men and women..." - and then journeyed on to Kapunda and other rural settlements. Despite a period of illness, he persevered with his missionary task and was before long joined by other volunteers from home. Within ten years the Bible Christian Connexion had more than 1,000 members in the colony, and had erected 37 chapels as well as establishing numerous meeting places. South Australia had become the centre of Bible Christian activity in the continent, in much the same way that Cornwall was the centre at home.

The Primitive Methodists, too, found support amongst the Burra miners. In September 1849 they opened their chapel at Kooringa, and by 1859 they had spread their influence to other mining settlements in the colony. In that year Captain William Arundel Paynter opened a Sunday School in his cottage at Wheal Ellen, near Strathalbyn, writing to his wife Sophia in Cornwall: "I can tell you there has been great revivals here and I thank the Lord he has revived his work in my soul ..." The Wesleyans also made an impact in the mining camps, Captain Morcom initiating services at Montacute in the mid-1840s and Captain Richard Rodda organising the opening of the Wesleyan chapel at Penrice in February 1855.

The first Wesleyan minister had visited the Burra as early as 1845, and in 1847 the Rev Daniel Draper wrote to the Wesleyan authorities in Britain:

I have ... the pleasure to inform you that a new Wesleyan-Methodist chapel has been opened at the famed Burra Burra mine, one hundred miles north of Adelaide... (the) population... is fast increasing... many of them are from Cornwall, a considerable number of whom were members of our society at home...

The following year Draper urged the Wesleyan Conference to send further missionaries, in response to "... the number of persons who come out from our societies and congregations in Cornwall, and are like sheep without a shepherd ..." Draper and his colleagues were hard pushed to service the wants of the rapidly expanding colony, but, like the other Methodist denominations, were remarkably successful in "civilising" the Burra.

The Methodists had boundless sympathy for all unfortunates - like poor Luke Teddy, a Cornish miner permanently blinded in a Burra prize fight, whom they took under their wing - but they also imposed a strict discipline upon their members. In August 1852, for example, a meeting of Kooringa Wesleyan class-leaders was held to -

... investigate a serious charge which had been preferred against Brother Adams by Sister Nicholas. The nature of which is as follows. Some weeks ago she heard the said brother, making use of the following language to his eldest son, at his back door, while in a rage of passion, "Go along you Bugger." Sister Nicholas mentioned the above to Brother Pascoe, by whom it was told to others. She never once named the circumstance to the Brother charged.

Sister Nicholas refused to attend the meeting, Adams junior said his father's word had been "hosher," and thus Brother Adams was aquitted and his accuser censured for having "... violated the laws of Scripture and Methodism..." by spreading false stories. Although strict, Methodist discipline was always applied with a keen sense of justice.

In addition to the Methodist sects, the Protestant Nonconformists at work amongst the miners at early Burra Burra included the Church of Christ (or "Scotch Baptists"). When Saltash-born Philip Santo moved to the Burra in 1849 to become Clerk of the Works, SAMA granted him land on which to build a chapel. It was said that "... Mr Santo was ... liked ... because his sermons were shorter and simpler and much more pathetic. He often shed tears which would run down his cheeks and have to be caught in his pocket handkerchief." One of Santo's assistants was George Pearce, a Cornish miner, who proved a gifted lay preacher. Only roughly educated, he was uncertain of his Biblical allusions, declaring that at the Second Coming "Then shall that glorious scripture be fulfilled - Jack's as good as his master!"

The early Nonconformists did not have it all their own way, however. Not only did they have to fight a political battle against elements of the

Adelaide establishment to preserve their religious freedom, but they had also a tackle the effects of the Victorian Gold Rush in the early 1850s. Their numerical strength was eroded by the wholesale departure of Cornish miners for Victoria, while a spiritual threat was posed by the new obsession with gold and wordly wealth. The Rev. James Rowe wrote that "For some time ... we had good congregations and much blessings at the Burra and Kapunda ... but the discovery of gold in Victoria the year after the beginning of our work shattered everything." The Primitive Methodists complained that the Rush had "... exerted a withering effect on many of our societies and congregations...," and the Rev. Daniel Draper added in July 1852 that "You will not be surprised to hear that these circumstances are by no means favourable to the spiritual interests of the people. The amount of worldliness is extreme; and our best people are lamenting the deadness which is induced..."

But it was not all bad news. A party of miners in Victoria sent £56 in 1852 to swell the funds of the Kooringa Wesleyan chapel, while Brothers Halse and Blamey - two Bible Christians from the Burra - "... promised the Lord five percent of all the gold he may give them." And when the men returned from the Rush, many rejoicing in their success, there was at the Burra a "Great Revival." By the end of 1853 James Rowe could write that "The cause here was recovered from the shock it received from the discovery of the Victorian Gold Fields," and in 1854 the Burra Wesleyans were recording "... conversions, the recovery of backsliders, and the prosperous state of the Sunday-schools..." In the same way, the Primitive Methodists could note that "... the mission has been recovering from the severe shock... more members from Cornwall have gone there." The diary of the Rev. John G. Wright, the Primitive Methodist minister at Kooringa, lends many insights into the Burra revival. For instance, there was the occasion when one "... morning I was called to visit a poor man who had been for days under conviction of sin. When I entered the house he fell on his knees with horrific groans. For one hour he was in the greater agony." And on other occasion, he

Attended a prayer meeting at Kooringa. A man by the name of F. Cock was brought to God. He had been notorious for crime, had took his wife by the hair of the head and swung her round the room. When made happy he ran round the house praising God. A poor man by the name of Jenkin had not long before got his jaw broken, he got so happy he praised the Lord shouting "I will praise the Lord ... though I have a broken chack (cheek)."

The late 1850s were indeed the great days of Burra Methodism. The Sunday Schools flourished and the district was alive with self-help bodies such as the "Wesleyan Mutual Improvement Society" and the "Redruth

82

Total Abstinence Society." In the 1860s the Methodists fared less well, for a great many families left the area to go to the newly-discovered mines at Moonta and Wallaroo. Nevertheless, the Bible Christians were fortunate in that their minister at the Burra, from 1862, was the Camborne-born Rev. Joseph Hancock - a man of tireless enthusiasm who was able to maintain and consolidate the Bible Christian faith at the Burra during the difficult decade of the "sixties."

At Kapunda, of course, the Methodists had also been hard at work. But Kapunda, even in the earliest days, never seemed to have quite as bad a reputation as that aquired by the Burra. Indeed, social conditions generally appear to have been never as poor as those endured by settlers in the Burra district. In 1846 Francis Dutton wrote that,

... several rows of substantial cottages, on a uniform plan, are already erected... The miners having their families now living with them, are happy and contented, and are not continually wanting to go to town (Adelaide) as they formerly did ... A chapel, which will also serve as a school-house, is by this time completed.

Life was not always as rosy at Kapunda as Dutton might have us believe, and one particular danger to life and limb was the mine itself. Open shafts and moving machinery were threats to the unwary, especially playing children, and the miners' cottages were constructed along-side the mine - thus giving easy access to the area. One of the children's favourite dares was to steal a ride on the balance-bob which was situated at the shaft-mouth and had an up-and-down rocking motion. The Kapunda miners became very angry when they found anyone riding the bob, for they knew it was a dangerous practice. In Cornwall there was an old rhyme in which a balance-bob "Scat the old man back in the shaft," and in fact at the Burra in 1875 Captain Sanders' young son John was "... frightfully crushed in the lower part of the body ..." when he fell under a bob during a game of chase.

The miners actually working underground - at Kapunda, Burra, or any other colonial mine - were in some physical danger. But there were no major disasters, and on the whole South Australian mines were considerably safer than their Cornish counterparts. Indeed, between 1866 and 1900 there were no more than 83 deaths in the very deep and extensive northern Yorke Peninsula mines. And in marked contrast to many overseas mines, the South Australian workings were generally considered healthy - with deaths from "miners' complaint" (a general term covering all lung disorders from consumption to phthisis and silicosis) being remarkably infrequent.

General health conditions in the Moonta Mines and Wallaroo Mines areas, however, were little short of appalling, and, whilst miners' pay may have been far higher than wages in Cornwall, actual living conditions could be as bad - and on occasions far worse - than at home. Overcrowding and the make-shift nature of the dwellings were the roots of the problem for, while the government laid out townships at Moonta, Wallaroo and Kadina, the majority of the miners preferred to build their own homes on the actual mineral leaves - thus forming the settlements of Wallaroo Mines, Moonta Mines, East Moonta, Yelta, Cross Roads, Newtown, and the several smaller suburbs. Wallaroo was on the coast, and the entire district was contained in or around a tract of land which formed an inverted isosceles triangle with sides (Moonta-Wallaroo and Moonta-Kadina) measuring some ten miles and a base (Wallaroo-Kadina) of six miles. Physically, the triangle became a curious mixture of mines, small farms, several concentrated settlements, and numerous scattered cottages, together with the three commercial centres, the skyline dominated by engine-houses and Methodist chapels. This random, partly urban/partly rural, environment was strongly reminiscent of Cornwall - especially of the ribbon development of the Camborne-Redruth area.

In 1875 the Rev. W.H. Hosken, describing the mineral leaves, wrote that "The mines as they are called, make a queer place. No streets, everyone builds his little cottages just as fancy leads him." And Anthony Trollope, after his visit to the district in the early 1870s, remarked that the miners "... have built habitations for themselves round the very mouths of the shafts, and in this way ... vast villages have sprung up - consisting of groups of low cottages, clustring together..." Many housewives, inspired by the Methodist committment to self-improvement, did their best to keep their cottages homely, clean, and tidy. But all too often there were also cases of extreme destitution and disgusting squalor. Despite the existence of a rather harsh Poor Relief system (which forced widowed mothers to take in laundry or sewing), there were still such appalling examples as the Sparnon family at East Moonta who, had it not been for the help of their neighbours, "... must have died from starvation and cold...," the father being "... a perfect wreck of humanity, with an almost heart-broken wife and seven small children..."

Perhaps even more alarming than poverty were the insanitary conditions that were prevalent on the mineral leases. In 1877 one horrified observer complained that,

... many of the residents do not study their neighbours' interests, and are quite unscrupulous as to where they throw their soapsuds and filth, and if the erecting of a pigstye or closet close to their neighbour's back door will cause an annoyance, they are but too highly gratified. You may oftimes see dead cats, dogs, fowls & C lying about the streets in a state of putrefation.

There were also many instances of bad overcrowding, such as in 1884 when William Bennetta of Wallaroo Mines was found to be living in a small cottage occupied by 12 persons, seven of whom slept in one room measuring 12ft x 10ft, the others sharing another nine ft square. Not surprisingly, William had been suffering from fever for six months. The situation in the mineral leases was exacerbated by the weather - both directly and indirectly - through the effect of drought and the primitive methods of water conservation employed by the miners. In the dry season of 1864/65 it did not rain on the Peninsula for six months, water then selling for 6d a bucket. 1877 was a bad year, and in 1879 the Government had to cart emergency water supplies to the area. In an effort to collect all rain water, tanks were sunk into the ground by the miners, or attached to their cottages. Not surprisingly, though, these tanks were soon polluted and became a breeding ground for all sorts of bacteria.

There were, therefore, periodic outbreaks of disease, which swept the mineral settlements, decimating the infant population, the greatest threat being typhoid (or "colonial fever" and "black measles" as it was also known). The district was rarely free of the fever, but the worst period was 1873 to 1875. During 1873 there were 327 burials in Moonta cemetary, mostly typhoid victims, with the *Register* noting that during June there were up to four funerals a day. The Central Board of Health, in distant Adelaide, became alive to the problem during 1874, urging the local mining companies to improve conditions on the leases and at the same time applying more stringent hygiene regulations. There were some improvements, but still in 1884 the Central Board had to threaten legal action against the Moonta company in consequence of the sub-standard housing on its land. By now, however, the mining companies had become aware of the extent of the problem and had themselves appointed local health authorities to police conditions on the mineral leases.

Perhaps surprisingly, northern Yorke Peninsula never suffered from a reputation similar to that of early Burra, although in 1877 one observer wrote of "... the great immorality existing on the Mines." He claimed that in an area "... renowned for its Christianity, you may see sights and hear sounds that would shock the modesty of any right-feeling human being...," and added that young ladies were "... almost afraid to go to Chapel..." for fear of being sworn at and hooted after. But generally, all the evidence suggests that the mineral lease dwellers were a peaceable and responsible people, and, for a community of some 25,000 souls, there was remarkably little crime. As at the Burra, the Methodists were a tempering and guiding influence for the local populace.

Indeed, it was in Moonta and environs - the northern Yorke Peninsula mining district - that Cornish Methodism was most deeply ingrained. The first-ever service in the area had been held in a tent at Wallaroo Mines in 1859, and at Moonta the first service was preached by James Bennetts, from Camborne, in a carpenter's bough-shed. But it was not long before the Peninsula had both ordained clergy and purpose-built chapels. In 1875 there were no less than 14 Methodist chapels in and around Moonta, with 21 within the whole Peninsula mining district. When the Rev. W.H. Hosken, a Bible Christian minister from Victoria, visited the Peninsula in January 1875 he was "... persuaded that I have never seen a finer field of labour anywhere in our Connexion..." In addition to the numerous Cornish local preachers and class-leaders, many of the Peninsula clergy also hailed from Cornwall. There was the Rev. Joshua Foster, who had worked in the Kilkhampton Circuit before volunteering for Australia in 1857, and the Rev. Charles Tresise from St. Erth. There was also Henry Pope, born in Helston in 1844, and S. Trethewie Withington - "... a minister long and well known in Cornwall both as a preacher and a writer of eminence."

Not surprisingly, local Methodism had a strongly "Cornish" flavour. The miners were "... fervid, strong in their feelings, love or hate, passionate if you will ... Those were days when preachers spoke of hell with an absolute belief in it as the abode of damned souls..." The Revivals, as elsewhere in the colony, were important events for the Cornish - opportunities to save backsliders and to win new conversions. There was a particularly impressive Revival in 1874-75. By then the Rev. John G. Wright had moved to Moonta, and as before his diaries are especially illuminating. He wrote: "There is grand work doing in all the Chapels." At Wallaroo Mines "Many were shouting and weeping," and at Moonta there were "Souls saved every night... How grand to see souls cast themselves on the altar. Go when you please, night or day, you will hear people singing God's praise."

Another important part of chapel life on the Peninsula were the Sunday School Anniversaries or "tea-treats." The children would form-up by class at the chapel and then, with banners flying and brass-bands playing, march about the streets for sometimes an hour or more. On their return to the chapel they would each be given buns and cakes, while the grown-ups sat down to tea. In the evening a meeting would be held in the chapel hall, when the reports and accounts would be presented to the chapel members. The Sunday Schools were also a vehicle of the Methodist self-improvement ethos, as were the various mutual improvement societies, and played an important educational role before the coming of State-run schooling. Indeed, several local Cornishmen established their own private day schools on land donated by the mining companies, while others were later involved in the launching of the State-education scheme.

In the same way, the Methodist "pre-occupation" with "improvement" led to the emergence of a vigorous Temperence movement on the Peninsula. Although only the most extreme objected to the moderate drinking of "swanky" and other beers, regarded by many as a wholesome working-man's beveridge, all Methodists attacked the horrors of "colonial brandy" and warned of the abuse of spirits and wines. They were especially concerned about the social effects of hard-drinking, and the "Bands of Hope" would often hold debates, inviting as guest speakers well-known local Temperence advocates such as Reuben Gill and W.J. (later Sir William) Sowden.

The Rechabite Lodges were amongst the most powerful of the Temperence bodies on the Peninsula, but of particular interest were the local Orange Lodges - who were not so much anti-drink as anti-Catholic. Their members were Cornishmen, mostly Methodists and all nonconformists, and they strongly opposed the building of Roman Catholic churches in the area. When finally one was opened at Moonta a violent confrontation between the Orangemen and local Catholics (principally Irishmen) was only narrowly avoided, and on the 12th July every year the Peninsula towns resounded to the stamping of feet, the beating of drums, and the sound of banners rattling in the wind, as the Lodges celebrated the Boyne anniversary.

Of course, Methodist chapel services were deliberately plain and simple affairs, and even Anglican ceremony was dismissed as "Papist." But, paradoxically, one of the most elaborate and atmosphere-laced religious events to be found anywhere in South Australia was the Peninsula funeral service. In 1902 the Rev. W.F. James, from Truro, recalled the unique spectacle of the old-time Cornish burial ceremony, describing funerals he had known at Chacewater and adding that he had "... never heard anything like it, save at Moonta." The sight of several hundred mourners, all dressed in black, accompanying the coffin to its final resting place, with the dirge-like hymns echoing hauntingly across the bleak mineral leases, was not something easily forgotten.

Another facet of religious life on northern Yorke Peninsula was the role of the mining captains. Not all fitted the stereotype of the up-right miner-preacher, but there were many who did. At Moonta, the Wesleyan chapel was always known as the "bosses' chapel" - the local preachers including the Captains Malachi Deeble, Christopher Faull, and Henry Richard Hancock. A mine worker looking for promotion had to be regular in his attendence at chapel, and one of the many "Cap'n 'Ancock" yarns asserts that a newcomer to Moonta Mines secured a job on the basis of his comment "... that was a lovely prayer you made at the meeting last night, Mr. Hancock." Similarly, another new employee at the mines went underground with Captain Hayes to be shown "the ropes." But Hayes... "was a local preacher and on the strength of my father being a parson we spent most of the time sitting in a stope while he told me of sermons he had read, sermons he had heard, and (mostly) sermons he had preached."

Over the years the Peninsula became established as perhaps the most important Methodist community in South Australia, carrying even more weight than Adelaide. Thus it was that the locality had a major "say" in policy decisions affecting the development of the three main denominations in the colony. During the 1870s, when there was an urgent need for new missionaries to minister to the rapidly expanding rural areas, it was a Moontaite - the Rev. John Thorne - who was sent home to address the Bible Christian Conference of 1876, held in St. Austell. He, incidently, was quite successful - the Revs. Richard Carlyon Yeoman, W.F. Jones and William Thomas Penrose (from

St. Buryan) volunteering shortly after for South Australia.

The Peninsula also featured significantly in the debate concerning Methodist Union, which was topical for more than 20 years at the end of the century, when rural contraction following the droughts and depression of the 1880s combined with the wholesale departure of Cornish miners for Broken Hill (and later Kalgoorlie) to undermine the independent viability of the Methodist denominations. In 1881 a Bible Christian report hinted at the necessity of Union if Methodism was to survive as a dominant voice, but many Bible Christians on the Peninsula were reluctant to have the specifically Cornish nature of their sect submerged in amalgamation with the other sects. But the Rev. James Rowe, a man of tremendous standing in Methodist circles in the colony, advocated Union and he was successful in bringing the Bible Christians closer to the Primitive Methodists and Wesleyans. In 1876 the Bible Christian Conference at St. Austell had given permission for a separate South Australian Conference to be set-up, and in 1881 this new body declared that it was at least keen to "... confer with the representatives of the Methodist bodies..." In 1888 the Bible Christians and Primitive Methodists actually voted for union, but it was felt that the majority was not great enough to be considered a mandate on such an important matter. Finally, in 1896 all three denominations declared in favour of amalgamation. A united Methodist Conference was held in Adelaide in 1899, and the Bible Christian Conference at Penzance in 1900 was the last attended by South Australian delagates. Formal Union was achieved on January 1st of that year, the Moonta *People's Weekly* talking enthusiastically of "... the great idea of a united Methodist Australian people...," and the minister at Kapunda declaring:

... not only would the union be a great and useful factor in the colonies, but to the town of Kapunda. Let their motto be that good old Cornish one - "One and All" - and let them work together for the cause of God.

Chapter 4

THE RADICAL TRADITION

Recent research has confirmed both the distinctive nature of Cornish politics and the importance of Methodism in moulding its development. And the Liberal Party, which in Cornwall had been founded upon each of these elements, has been the main vehicle of what we may call the "Cornish Radical Tradition." This tradition emerged in Cornwall in the nineteenth-century and its principal elements were transplanted in South Australia, where they developed in a manner reflecting the Cornish background but at the same time changing in response to new, Australian conditions. This transplantation was at two levels - in the attitudes and dispositions of individual migrants, and in the political climate or culture created in areas of Cornish concentration such as Burra Burra and northern Yorke Peninsula. Four major manifestations of this transplantation can also be traced - a strong link between Methodist and political activity, a tendency towards liberal or social democratic sympathies, the development of trade unionism against the background of the tribute and tutwork systems of employment, and the evolution of an advanced Welfare programme in the larger mining companies.

In nineteenth-century Cornwall Methodism, with its anti-Anglican attitudes, represented a kind of social or religious radicalism, and it was inevitable that as political radicalism emerged so the Methodists would be attracted to its ranks. Many saw their support for political radicalism as merely a practical extension of their Christian convictions. Christianity taught compassion, concern for the needy, the equality of men before God, a contempt for riches; and politics was the means by which God's Word was put into action.

In mainland Britain, it was principally the Celtic areas - Wales, Scotland, and Cornwall - which proved to be the most non-conformist and the most non-conservative. In Cornwall in particular, the link between Methodism and Liberalism has always been an obvious one, noted by most Cornish and Methodist historians, with the Bible Christians being the most radical of the denominations. Cornish Methodists were quick to involve themselves in local politics, and it is interesting to note that in the first Cornwall County Council, elected in 1889, there were 25 Methodists of whom 18 were Liberals, three extreme Radicals, and only four Conservatives. Local Methodist organisations also tried to influence legislation passed at Westminster, so that in 1902, for example, the Kilkhampton Bible Christian Circuit added its voice to the widespread Liberal and non-conformist opposition to the Conservatives' Education Bill. The same Circuit's members also articulated their hostility to Conservative and Anglican forces at a more

89

theoretical level, declaring that they stood "... by the privileges of civil and religious liberty won for them by their forefathers at a great cost and handed down to them not simply to enjoy, but to establish and extend."

The Methodist connection apart, the Liberals enjoyed growing support from the Cornish people after the Great Reform Act of 1832. Indeed, the Liberal Party became firmly and completely established in Cornwall so that, for example, between 1885 and 1910 all six Cornish constituencies returned Liberal candidates at every election. Often seats were contested, not by opposing Liberal and Conservative candidates, but by several Liberals of slightly differing political complexion. Thus one memorable election was the "... intense and bitter... contest between Whig and Radical in 1885..." when the Independent Liberal, C.A.V. Conybeare, defeated the official Liberal candidate, A. Pendarves Vivian, in the Camborne constituency. The major Cornish newspaper in those days, the *West Briton,* was also strongly Radical Liberal in its outlook.

Unlike the Welsh experience, however, this non-conformist Radicalism did not prove to be the precursor of a deeply-ingrained Labour movement. The Labour Party was slow to develop in Cornwall, and has never been strong, and those local Labour politicians who have achieved a measure of success have been on the right-wing of their party, sharing with the Cornish a strong suspicion of the Marxian and intellectual Left.

The farming influence in Cornwall accounts in part for the non-socialist nature of Cornish Radicalism, but most historians attribute the relative weakness of the Labour Party to the weakness of trade unionism, which is in turn attributed to the economic organisation of Cornish mining in the last century. Unlike the usual employer/employee relationship to be found in most industries, which led to the antagonisms which were the parents of trade unionism, there were instead the Cornish "tribute" and "tutwork" systems of employment in which part of the entrepreneurial function was performed by the miner himself.

In the tribute system (which was soon to find its way across the world, from California to Victoria), individual sections of the mine or "pitches" were contracted out to individual miners or groups of miners ("pares") as a result of open bidding. Before this bidding, which occurred on "survey day," each pitch would be inspected by a captain who would estimate the value of the ore it contained. The company would then offer each pitch at "captain's prices." For a rich section of ground containing high-grade ore, the "captain's price" could be as low as just a few shillings in the pound, meaning that for each pound of the value of ore raised the tributer would receive a couple of shillings. But for a low-grade pitch the "captain's price" might be as high as 15 shillings in the pound, an incentive for the tributer to work indifferent ground.

Sometimes tribute pitches were in fact let at "captain's prices," but more often there was considerable downward bidding between rival tributers

- especially for attractive sections of ground. The effect of such bidding, of course, was to reduce the tributer's share of the ore value to well below "captain's prices," thus lowering his final income. The tributers were expected to provide their own candles, powder, and materials - which they obtained through the company - and at the next survey day (they were usually monthly) they would be presented with a "bal-bill" showing the cost of materials deducted from their payment. If a miner had been forced to draw "subsist" (an advance on earnings) during the period between survey days, then it was possible that when he received his bal-bill he would (after the various deductions) be entitled to no further income or indeed be in the position of owing the company money.

The tribute system, nevertheless, gave the miner an opportunity to make use of his enterprise and skill in a way that was impossible with ordinary day-labour, and was thus generally well-liked by the miners. By working hard, a tributer could amass a large amount of ore and thereby increase his income; and occasionally he might locate a rich pocket of ore (a "sturt") in otherwise unpromising ground and so earn a vast amount during that particular "take."

But despite the great advantages, the tribute system weighed often against the miner. In addition to the financial problems noted above, it was sometimes the case that the lode in an apparently rich section of ground would suddenly give out - leaving the miner with poor grade ore but with a contract which allowed him only a small percentage of the value of ore raised. In such situations, tributers were sometimes forced to steal from their colleagues in order to make a reasonable income, and at other times tributers saved money by not erecting adequate (but expensive) timbering to support their workings. And the greatest feature of the tribute system - along with the associated tutwork system in which contracts were concerned with the volume of ground mined rather than ore value - was that it tended to set miner against miner, forcing the tributer to compete against his fellows at periodic intervals for pitches offered to the lowest bidder. This, more than anything, helped to frustrate the growth of trade unionism in the industry and retard the development of a wider Labour movement in Cornwall as a whole.

However, although trade unions were slow to develop in Cornwall, it is interesting to note that when they did finally emerge they bore all the marks of the Cornish non-conformist Radical Tradition. The trade union activists were often also officials in the Methodist chapels, class-leaders and local preachers. They were better-educated, articulate, experienced in the organising and leadership of men. But more than this, their industrial activity - like their support for the Liberals - was for them the practical implementation of their Christian beliefs. There were, indeed, some similarities between Methodism and trade unionism. There was common emphasis on fellowship, on mutual support and improvement, on solidarity, and there were even similarities in organisation and terminology - the members of both Methodist classes and trade union branches, for example, referred to each other as "Brother," a term of egalitarian significance in both movements.

Another feature of the unusual organisation of the Cornish mining industry was the development of a Welfare system which, by comparison with conditions prevailing elsewhere, was remarkably advanced and sophisticated. At a number of mines there was a "bal-surgeon" to give on-the-spot attention to the miners, and every worker contributed to a "Club & Doctor Fund" which paid a miner's medical bills and afforded him subsistence allowance when illness prevented him from working. At some of the larger mines there was even a mine barber. Being a reaction to local industrial conditions, and to a local political culture and Cornish social attitudes, the creation of these Welfare systems was in a sense an integral part of the Cornish Radical Tradition, and deserves to be treated as such.

As noted in earlier chapters, there were a considerable number of Cornish non-conformists in early South Australia, many attracted by the colony's "liberal-dissenting" atmosphere, of whom several became involved in Adelaide's political life. John Stephens, the South Australian propagandist, came out to manage the liberal *Register* newspaper, and Penzance-born George Marsden Waterhouse (for a short time Premier of the colony) was a leading early liberal politician. Self-government in 1857 opened the way for other Cornish colonists to become involved in Parliamentary affairs. Philip Santo, for a time Clerk of the Works at the Burra Burra mine, was described as a "liberal nonconformist" by his contempories, and sat in Parliament continuously from 1860 to 1882 - five years as Commissioner of Public Works. He was also successful in persuading his two Cornish colleagues in the Church of Christ, George Pearce and James Crabb Vercoe, to enter the Adelaide Parliament.

Of particular importance was James Penn Boucaut, who although not a miner is of some interest here for he represented the Burra district in Parliament for a time, while also lending invaluable advice to the early trade unionists at Moonta. Boucaut was born in Mylor in 1831, and although his father was a Guernseyman (hence his French-sounding name) he was proud to describe himself as a Cornish liberal: "I have always stood by my counties (sic) motto 'One and All' and contemn (sic) and despise mere money shoddy aristocracy." He was first encouraged to enter politics by "... friends who knew my democratic sympathies...," and was elected to the Adelaide Parliament in 1861 - gaining invaluable support from the working class "Political Association." He formed a lasting political alliance with Philip Santo, and although he was defeated in the elections of 1862 he was again in Parliament from 1865, and from 1868 represented the seat of Burra Burra.

In the period 1872-75 Boucaut emerged as clear leader of the liberal Opposition groups in Parliament, finally being elected Premier. He introduced a grand scheme for the rapid expansion of the colony, with his

92

projected rolling back of the agricultural frontier to be accompanied by the building of 13 new railways to penetrate the far-flung corners of the colony. Although this sweeping, not to mention costly, plan met with stiff opposition from the conservative upper house, which temporarily cost him the Premiership, Boucaut had the satisfaction of seeing almost all aspects of his scheme come finally to fruition.

Perhaps even more interesting than this political career, for those wishing to gain insights into his general disposition, are his private views as articulated in his intriguing correspondence with J. McArthur, one-time secretary of the Cornish miners' union at Moonta. Boucaut was always a great friend of the miners, he called himself "Cousin Jackey," and he supported the Moonta and Wallaroo men through the strikes of 1864 and 1874. It was in 1874, after the "Great Strike" on the Peninsula, that McArthur first wrote to Boucaut, asking for his advise on the newly-formed union's political policies. Boucaut's response was warm and enthusiastic, explaining that "... our Legislation and system of Government studies entirely too much the interests of capital. I am very glad to see such Unions as yours established..." To impress his sincerety he added that, "I wish you to believe that I do not profess liberal sentiments in order to gain power. I should have far more power if I were to hold contrary sentiments." Much to McArthur's relief, no doubt, he also added "I differ from those who think that the Union should be disassociated from politics." This was because "capital is true to itself while labour is too often true neither to itself nor its friends... Labour was once enslaved. It now demands liberation..." Looking to the future, Boucaut declared that "There is no reason why the man whose industry makes the article should not look forward to the time when he will be on a perfect equality in every respect with the man whose capital aids him in doing so" - a view which was reformist rather than revolutionary, social democratic rather than socialist.

Boucaut criticised the embryonic Moonta Miners' Association because if exercised its power "... by fits and starts which is bad both for the country and yourselves..." and because it "... had no settled principles and no cohesion..." He also urged the men not to be divided by sectarian feeling: "Be true to yourselves," he argued, " - if a Wesleyan vote against a Bible Christian because he is of a rival church both suffer." He urged the unionists to establish links with like-minded bodies in Adelaide, and argued that internally the union should concentrate on education, whilst politically it ought to agitate for payment of Parliamentary members as a prelude to nominating their own Labour candidates - "... everything comes back to payment of members and education. These are your two great necessities."

From the union's point of view, Boucaut's advice was sound, and it is interesting that after 1874 the miners did become more efficiently organised and more politically aware. It would, of course, be naive to attribute this entirely to Boucaut's influence and instead it is necessary to refer to the changing nature of the miners' industrial disputes in the colony. In particular, it is necessary to examine the changing attitudes to the tribute and tutwork systems, and to remember the development of trade unionism elsewhere in South Australia (and at Broken Hill).

The tribute and tutwork systems of employment were adopted in South Australia in the earliest days. Robert Nicholls, a Cornishman and the first miner at Kapunda, was employed *on tribute*; and at the Burra the systems were in use from the very beginning. In 1846, in the Burra's first full year of production, the *South Australian News* recorded that,

Miners are no longer satisfied to work "on owner's account"
for less than 30s per week, especially at any distance from town;
but the prevalent wish of this class is to be placed on "tribute"
or "tutwork," which is natural enough when they know that
some of their brethren are earning such splendid remuneration
at the Burra Burra Mines.

The success of the systems at the Burra did indeed induce other mines to introduce them at their workings, and at the end of 1846 the *South Australian News* could again note that "... the Cornish system of 'tutwork' and 'tribute,' or one of the contracts suited to the circumstances, will doubtless prevail..." in the colonial industry. The miners themselves were clearly pleased with this situation, a number commenting favourably, in letters written home, on the introduction of the systems.

And when industrial conflict broke out at the Burra in September 1848, the bone of contention was not the system of employment - with which the miners were well pleased - but rather the results of the assaying performed by the South Australian Mining Association (SAMA) when determining the value of the tributers' ore. Under the SAMA's rules, the value of ore in a particular pitch would be asscertained by comparing three different assays - that of the tributer who had worked the pitch or intended to work it in the future, that of Mr. Thomas Burr (the then Superintendent), and finally that of the mine's chemist. The Directors had become critical of Burr's assays, which they considered too much in the miners' favour, and were inclined to accept the more conservative assays of their chemist. They also accused the tributers - or rather their representative, Hosken, who actually performed most of their assays - of making the assays "... greater than they really are."

Burr lost his job, and the miners reacted angrily to the accusations by mixing together their ore as a protest, and by striking. They were quick to add other grievances, too, some tributers complaining that their ore had been left lying around for months before SAMA settled-up, and others pointing to the high level of bal-bill charges.

The disturbance caused some alarm in the colony, one correspondent to the *Register* noting gravely that 1848 had been a year of European

Revolution. In anticipation of violence, a squad of police troopers was sent to the Burra, but they were not needed - even though "black-legs" were pinioned to wheelbarrows and carried shoulder-high through the town, "... exposing them to the gaze and ridicule of 1800 laughing souls." The call for police, however, was indicative of SAMA's general attitude - it was determined to keep tight control over the miners and tended to over-react in industrial situations. Henry Ayers (the SAMA secretary - not the most diplomatic of men) had a particularly haughty disposition and a somewhat hostile attitude towards the miners, together with an almost fanatical concern to maximise profit and output and to minimise cost. In April 1846 Mr. Boswarva, the mine clerk, had been reprimanded for being too familiar with the miners (his fellow Cornishmen), while in the January Ayers had explained sternly to the Superintendent that "You will please understand that the Labourers are to have £1 per week only... and if the men choose to leave I can soon replace them by others to any extent you may require."

The antipathy and insensitivity displayed by Ayers and the Directors was clearly a factor in prolonging the strike. The *Register* noted "... the discourteous line of conduct on the part of the Board of Management ...," and SAMA foolishly aggravated the situation by announcing a reduction of wages for day-labourers - a move which provoked a memorial from the men demanding a full restoration of rates and privileges for all "... miners, labourers, carriers on the mine, whim boys, and ore-pickey boys."

SAMA reacted, again characteristically, by offering at the next survey-day only seven tutwork bargains and 14 tribute pitches (instead of the usual 70 or 80) with the direction that other miners would have to work on a daily-wage basis - a move calculated to punish the men who would resent the resultant loss of independence. By this time, though, the strike was petering out - some of the ringleaders having been ordered from SAMA's cottages and forced to leave the Burra, the others through economic necessity having to return to work. Inevitably, there were further recriminations. Some, like Joe Trevean, A. Penna, Thomas Cocking, M. Rogers, and T. Polkinghorne, were re-employed on the understanding that they behaved themselves while others, such as Messrs. Mitchell and Moyle, were told that they might be taken on at a later date, and still others - including Messrs. Bosance, Hosken, and Stephens - were refused work and given notice to quit the SAMA cottages. Wage rates, needless to say, were not improved, Captain James Trewartha writing an indignant letter to the press claiming that the position of the South Australian miner was "... not a whit better than that of the Cornish miner."

The Burra strike, then, was a clumsy affair - badly handled by both employees and employers - and arose not through a deep-seeted hostility to the system of employment on the part of the miners, but as a result of discrepancies in assaying and the general insensitivity of SAMA. In no way was it an attempt by the Cornishmen to establish a union or to combine consistently against their employers. In short, the strike was a spontaneous and ill-organised reaction to a particular situation rather than a strategy to alter the basic relationship between employee and employer. Precisely the same could be said about the strike at Moonta and Wallaroo in 1864. It did not represent the emergence of "organised labour" on northern Yorke Peninsula, but was in fact an expression of disatisfaction with the managerial regimes of Eneder and William Warmington, and a reaction against the two companies' *failure* to introduce the tribute system at their workings.

The Warmington brothers - James, Eneder, and William - were Cornish mine captains with experience both at home and in North America, and were appointed to positions of responsibility on the Peninsula in the early 1860s. But their sojourn there was never a happy one. James was dismissed as Chief Captain of Moonta Mines in October 1862 on account of "misconduct," and was replaced by his brother William. In the meantime, Eneder had been appointed Chief Captain at Wallaroo Mines. However, neither William nor Eneder was successful in establishing satisfactory relationships with their staff and workers. By May 1863 an atmosphere of hostility had grown up between William Warmington and Captain Osborne, a potentially difficult situation which had to be smoothed over by the Moonta company. And in March 1864 things came to a head when both the Moonta and Wallaroo miners came out on strike against the "two tyrants."

At Wallaroo Mines a group of timbermen who had been "unfairly" instructed to carry out work over the Easter break refused to do so, and were dismissed. This was the opportunity the men had been waiting for to express their disatisfaction with the Captain, and so the Wallaroo miners came out on strike. The Moonta men, looking for a similar opportunity to protest against the equally unpopular William Warmington, struck in sympathy with their Wallaroo brethren. Together, the two groups of miners began to rationalise and articulate their opposition to the Warmington brothers.

Both were accused of ill-treating their sub-ordinates: in particular, it was alleged that William had savagely flogged a young boy for a petty misdemeanour. The miners, under the leadership of Reuben Gill, Collingwood Kitto, John Lander, and a Mr. Knowles, compiled a memorial demanding the dismissal of the Warmingtons, and submitted it to the companies. At first the companies were unwilling to concede any ground, but under pressure from both the miners and the general public they began to relent - their first concession being to agree to an inquiry to investigate the behaviour of the brothers. And then suddenly Eneder announced his resignation from the Wallaroo Mines. This was due in part to his continuing ill-health (he died shortly after), but the miners saw it as an admittance of guilt. Sensing victory had been achieved, they agreed to return to work. The Moonta men also indicated their willingness to go back, under the temporary supervision of Captain East from the New Cornwall mine.

William Warmington, in the face of considerable public criticism and the almost total hostility of the miners, had little choice but to tender his resignation. The Moonta company felt that it had little evidence to condemn William, but still thought "... it would be expedient to accept the resignation." And so peace was restored at the mines, and the way opened for Captain Henry Richard Hancock to establish his rule at Moonta. The men themselves were satisfied with their triumph, and were equally pleased when the companies announced that the tribute and tutwork systems were to be introduced at Wallaroo and Moonta - this being the recommendation of Captains Prisk and Trestrail, who had been engaged during the strike to sound-out the mood of the miners.

The general effect of the 1864 strike on the Peninsula was not to create a rift between employers and employees, but in fact to draw the two sides closer together. The miners' venom had in any case been directed at the Warmingtons, and the companies had won their men's loyalty by acceding to their demands. This was why, perhaps, the rapid development of the Moonta and Wallaroo Mines in the years after 1864 was not punctuated by industrial conflict of any consequence. Certainly, it was not until a decade later, in 1874, that the mines were again brought to a standstill through strike action.

The 1874 "Great Strike," as it came to be known, has always been portrayed by Oswald Pryor and others as a rather light-hearted, almost jocular affair. But in fact the strike was especially nasty, creating for the first time an atmosphere of hostility between employee and employer on the Peninsula, and laying the foundations for a vigorous trade union movement at Wallaroo and Moonta. The event passed into the lore of local unionism - its ringleaders elevated to the level of heroes and martyrs - and an epic poem written to commemorate and immortalize the struggle, its melodramatic style adding further to the mystique which grew up around the strike:

> And soon they met at Elder's Shaft;
> The "Ring" was form'd, the glor'rous ring,
> Where Cousin Jack stands like a king;
> And freely to each thought gives vent,
> As to his brain each thought is sent,
> In eloquence that's all his own,
> States his opinions quick and clear,
> Nor will he yield to force or fear
> Nor will he back from anything
> He states when standing in the "Ring."

The initial cause of the strike was a decision of the Moonta company to reduce the general level of remuneration, notice of which was given on April 2nd. The men, reflecting upon a decade of phenomenol profits enjoyed by the company, demonstrated their indignance by striking, and were joined immediately by their Wallaroo colleagues who struck in sympathy. The solidarity displayed by the miners was almost complete, the few remaining

97

black-legs working the pump-engines being driven from their posts by the enraged miners' wives. The first reaction of the two companies was to stand their ground, complaining that the men had acted improperly by stopping the pumps. But later, having received a delegation from the miners and listened to their grievances, they agreed to restore the former wage rates in return for a return to work, and to give in future two months notice of any impending reduction. For the miners, this was a great victory and it was attended with impressive celebrations.

The real significance of the strike, however, was not the victory itself, but the new attitudes it created. Having witnessed the benefits of combination, the workers established and maintained an active trade union structure. Before 1874 there had been a half-hearted attempt to form a friendly society, but it was more concerned to establish a miners' co-operative store than to engage in industrial action, and few saw it as a precursor of a fully-fledged union movement. Some excitement was created in its ranks in July 1873 when there was a brief strike by miners at the Blinman, but by 1874 it was almost defunct. The "Great Strike," however, resurrected it as the vibrant "Moonta Miners' Association," a similar "Kadina Miners' Association" being formed at the same time for the Wallaroo men. At first both Associations were self-conscious and unsure of themselves: McArthur wrote to Boucaut for his advice, the Moonta and Wallaroo men squabbled over the rules, and two of the leaders - John Visick and John Prisk - quarrelled bitterly. However, despite the difficulties, the Associations welded themselves into a cohesive trade union movement which was able to test its strength by demanding a "closed shop" at the mines, and was by September 1874 formulating a clear political policy calling for such things as free education and payment of Parliamentary members.

Inevitably, there was a strong Methodist influence within the union, and the majority of union leaders were in fact Methodist local preachers. Of all the ring-leaders, none was more colourful than John Prisk, John Visick, and Reuben Gill, and it is interesting to note that each of these was intimately involved with practical Methodism. Prisk was secretary of the Moonta Miners' Association for many years. One newspaper piece referred to "... the Holy Land of Moonta under the able leadership of their modern Gideon, Mr. J. Prisk...," using these Biblical allusions to describe his qualities, and a local Bible Christian minister was pleased to note that Prisk was "... a lay preacher of the Gospel in connection with the Bible Christian Church..." John Visick (or Visack) came out from Kea in mid-western Cornwall in 1857 to work as an engineer on the Burra Burra mine. The son of a schoolmaster, he was a life-long radical, and was fiercely committed to the Wesleyan church - so much so, in fact, that when the Methodists moved towards unity in the 1890s he joined instead the Church of Christ as a mark of his protest!

Reuben Gill was even more of an individualist than Visick, and became known as "The Billy Bray of South Australia" - an indication of both his preaching style and the esteem in which he was held in local Methodist circles. He was a well-known Bible Christian local preacher, Rechabite, and teetotaller, having arrived in the colony circa 1850 to work first in the Burra mine. He was a much-liked personality and a favourite speaker at the miners' strike meetings. One commentator noted that,

Mr Gill's style of speaking was extraordinary. Jumping upon the platform as though propelled there by a catapult, he would jerk his head from side to side, and instantly let loose his eloquence at a tremendous rate.

Another remarked that,

His rough eloquence would fall from his lips in a rapid stream, and apt metaphor and racy extemporaneous rhyme follow each other with lightening - like rapidity, while the attention of his audience would remain enchained throughout his speech.

Because of his influence and charismatic appeal, Gill was ulitmately barred from employment in the Peninsula mines, and he ended his days in Adelaide - first as a mechanic, and then selling insurance. This "blacking" of Gill, indeed, was indicative of the hardening attitude towards activists on the part of the Moonta and Wallaroo companies, another important outcome of the 1874 strike. At the end of the conflict there were the inevitable recriminations. Three unionists who tried to generate further trouble were dismissed, and the Moonta company informed Captain Hancock of the "... advisability of getting rid of rowdy characters...," adding that it would "... not retain in their service men who attempt to interfere with the Management of the Mine." When men were made redundant for economic reasons, it was the union activists who were the first to be dismissed. McArthur, the author of the correspondence with Boucaut, was one of the first to go, and he wrote a pitiful letter to the Moonta Directors explaining the grave mistake he had made in joining the union and pleading to be reinstated. Rather uncharitably, Reuben Gill condemned him as a "traitor" and a "Judas."

The Moonta company had sought Cornish miners more or less exclusively to work their property. But the display of solidarity in 1874 frightened the Directors, and in an attempt to undermine the social cohesiveness of the Peninsula community they agreed in early 1875 to try to employ Irish and Scottish miners. And a few months later it was suggested that they obtain "... 200 Miners or Pitsinkers of other than Cornish nationality..."

In contrast to the miners' strikes of 1848 and 1864, the 1874 confrontation heralded a new era of industrial relations in the colony's mining industry. Although one frustrated unionist claimed in July 1878 that

"... Cornish men do not understand the true principles of Unionism and never will during the present generation," the miners became organised and active in a way that they had never been before. The great deference and respect the men showed towards the majority of their captains was eroded considerably, so much so that by 1879 John Prisk could declare that "If Captain Hancock knew how much the men spoke against him ... he would be ashamed to look the men in the face."

There was also apparent, after 1874, a changing attitude towards the tribute and tutwork systems of employment. At first this found expression in criticism of the *method* of setting tribute and tutwork contracts at the mines. Immediately after 1874 a particular "bone of contention" was the "five-week month" adhered to by the companies for the purpose of determining pay periods. During 1872 and 1874 miners in Cornwall had protested against the "five-week month" and had been successful in securing its abolition. Evidently the Wallaroo and Moonta men followed this example, and were spurred on by newcomers from home. One "New Chum" on the Peninsula in February 1877 wrote to his relatives in Cornwall, saying:

... no doubt you will be surprised when I tell you that the system of five weeks' pay which the miners not only in our mine but of almost every mine in Cornwall fought so valiently to crush is in operation in this far-off land.

But even more significant than this mounting criticism of the *practices* of tribute and tutwork was the growing disatisfaction with the *actual systems*. For despite the miners' initial enthusiasm for this contracting, the disadvantages of tribute and tutwork were as clear in South Australia as they were in Cornwall. In 1865, for example, the Moonta Directors had been forced to dismiss two tributers for ore-stealing, while just months later two more were sacked for the same reason. And the system did not improve with age - in 1889 two tutworkmen, Simmons and Trebilcock, were suspended for six months for failing to adequately timber their workings. It was inevitable, therefore, that as time went on the miners would come to appreciate the systems' inherent disadvantages.

In September 1879 came the first major sign of discontent with the systems and, perhaps in response to this, the companies replaced the practice of "open bidding" with "private tendering" - the miners now being required to compete for tribute pitches or tutwork bargains by tendering their offers to the company on slips of paper, rather like a secret ballot. However, this innovation made little difference for, as one observer wrote in April 1880,

Cornishmen are generally of an envious disposition, the characteristic principle being a dislike to see a neighbour advance a step ahead, and so with tendering. I have heard a remark passed when a miner has heard that another is tendering for the same job. "Ef 'ee d'get un 'ell have tew tender some law," and hence they all tender low and labor is brought below a nominal figure.

That this was so was demonstrated graphically in 1879, when tenders were invited for the sinking of four fathoms of a six-foot shaft at the East Moonta mine:

T. Williams £26 per fathom (accepted); S. Stephens £28; J. Angove £28 5s; J. Rowe £28 2s; R. Hancock £29; W. Pengelly & Co. £34; R. Gill £35; J. Bray £37; H. Stephens £38; J. Murrin and S. Samson £44 15s; Samuel Richards £90 for the first two fathoms and £100 for the next two fathoms.

Thus while one miner believed the job could only be undertaken economically at a price of £190, the contract went to a man who had tendered as low at £104. It is not surprising, then, that one miner should remark that "The fact is we miners cut each other ... and it is to be feared that our price is often far below the captain's price, which is cutting wages with a vengeance."

With a trade union structure now in existence, this growing opposition to tribute and tutwork contracting could be easily organised and articulated. There was a brief strike over the matter in 1884, and in 1886 there was further trouble following the abolition of "subsist" for long-term contracts (these had been introduced in 1877, and could span several months) and its replacement by a "percentage retained" system. Instead of applying for "subsist" when their funds ran low, the miners were now paid a weekly rate, the amount being calculated as a weekly average of the estimated total earnings over the contract period. However, a percentage of this weekly rate was retained by the companies in case the miners did not realise their estimated earnings, the retained money being paid out on the expiry of the contract. By March 1888 the percentage retained had reached 25 percent, and the miners - considering this far too high - came out on strike. The Directors agreed to reduce the percentage to 12.5 percent, but the men did not return to work until Captain Hancock had also promised them an increase in weekly rates. Later it was also decided, again in deference to union pressure, that a percentage would only be retained if the average weekly earnings were above £2 2s. 0d.

By this time, however, the miners had become heartily sick of the whole principle of contracting. In February and March of 1889 they submitted memorials to the Moonta company, at first seeking the abolition of contracting and then moderating their demands to "... requesting the Board's assent to a list of rules dealing with underground contracts." Both approaches were rejected out of hand, thus setting the stage for several years of hostility and conflict. During 1889 the local union associations became branches of the

"Amalgamated Miners' Association," an Australia-wide union based in Victoria, thus further enhancing their power and importance. This was part of a wider upsurge of unionism in the colony. The "United Trades and Labour Council" had been formed in Adelaide in 1884, and new unions were emerging while others were becoming more vocal. The rise of Broken Hill in the 1880s, too, gave Australia a new and vigorous centre of trade unionism, some of this vigour rubbing off on the Peninsula as a result of the strong links that were formed between the two areas. For many Moonta and Wallaroo men there came the realisation that their disatisfaction with the present system of employment could be voiced effectively only through a strong and active union. As W.G. Spence recalled,

The Cornish miner is generally a man who can do his share of grumbling, and frequently reckons he knows how to run a mine better than the manager, so when Unionism caught on they realised that many injustices might have been remedied years ago had they been organised and pulled together, instead of merely growling as individuals.

With the growing tension at Moonta and Wallaroo in 1889, a strike seemed inevitable. The Moonta Directors having first refused any changes in the employment system, the Wallaroo Directors then announced a general lowering of wages, the union branches having little choice but to threaten industrial action. The men stopped work on 13th May, but to try to smooth over the situation representatives from the "Amalgamated Miners' Association" (AMA) in Victoria travelled to the Peninsula. They were successful in winning a number of small concessions for their members, and in return explained to the men that the companies had demonstrated that the present economic situation would allow neither a rise in wages nor a major upheaval in the employment system. Grudgingly the men went back to work, but the agitation continued. New demands were made in the July and August, and in the December there was a brief strike by the moulders and fitters at Moonta. In the following February the union forced the companies to honour the "percentage retained" agreement, and in March 1890 500 men struck at Wallaroo Mines - staying out until they had won further concessions from the company. But the basic grievance - the system of employment - had not been removed. In July and August of 1891 negotiations about the system broke down, and so in the September the Moonta branch of the AMA declared an official strike.

This particularly bitter strike was characterised by the extreme hardship it caused. Miners all over Australia sent donations to the Peninsula, but still the AMA found it difficult to maintain strike payments. Finally, after 18 gruelling weeks, the strike was broken by desperate miners going to the mine office to plead for work. The AMA conceded defeat, and in return the company promised "substantial modifications" to the employment system. Nevertheless, despite this defeat, there was sporadic agitation against tribute and tutwork contracting throughout the 1890s. Partly in response to this, numerous adjustments were made to the methods of letting contracts - the most significant being in 1903 when the practice of "captain's prices" was replaced by a "sliding scale," based on changes in copper prices, in which the contractor participated directly in the company's profits.

The union itself survived the trauma of 1891, establishing itself further as an integral part of local community life. And the company's reaction to the consolidation of trade unionism on the Peninsula was not, as might be expected, one of increasing hostility and suspicion. For although the Directors opposed the threat of unionism itself, they also - quite paradoxically, and again as in Cornwall - developed for their workers a welfare system which was amongst the most advanced in Australia. This was initially a direct transplantation of the Cornish experience. The South Australian Mining Association had followed many aspects of Cornish mining practice at its Burra Burra mine, one facet of this being the creation of a welfare system. SAMA hired a bal-surgeon, introduced a Club & Doctor Fund, constructed a company hospital, built cottages for its employees, donated land for chapels, and paid for miners to bring out their relatives from Cornwall. When the Burra mire was first started - in 1845 - it was of course necessary for SAMA to meet many of the needs of the rapidly growing Burra community, for the mine was situated a hundred miles north of Adelaide in virgin bush, there being no established services or communications. This meant SAMA inevitably developed a paternalistic attitude. However, another impetus to the establishment of these policies was that they tended to increase the Association's control over its men - as evidenced in the 1848 strike when the ring-leaders were ordered from the SAMA cottages.

Be that as it may, the development of the Wallaroo and Moonta Mines in the 1860s was accompanied by the foundation of a similar welfare system on northern Yorke Peninsula, controlled through the "benevolent dictatorship" of Captain Henry Richard Hancock. Miners were given permission to construct their cottages on the mineral leases, other cottages were erected by the companies, a Club & Doctor Fund was set-up, contributions made towards local Institutes and societies, pickey-boys employed on the conditions that they attended school part-time, "old hands" found surface employment when they became too elderly to work underground, instruments provided for brass bands, land for chapels and schools donated, and health boards appointed. This system was developed to its high-point under H. Lipson Hancock, who succeeded his father in 1898, and became famous across Australia as the "Betterment Principle."

The success of the welfare system at Wallaroo and Moonta prevented the deterioration of industrial relations to the level of that, say, later experienced at Broken Hill. But, just as it did not prevent the growth of an enthusiastic union movement, so it did not retard the emergence of political radicalism. As noted above, early South Australia produced liberal Cornish Parliamentarians such as Santo and Boucaut, but there also developed in the colony's mining districts a strong tendency towards political radicalism which found expression, first of all, in electoral support for individual "liberal" of "progressive" candidates, and later in a committment to the United Labor Party. The ULP was in several respects different to its British counterpart, which never found much support amongst the Cornish miners at home, and was in some ways similar to the Liberal Party in Cornwall: radical but ultimately non-socialist, and drawing much of its moral strength from the Methodist Church. In the 1890s, when it first emerged, the ULP was known for its moderate, pragmatic approach - one writer in 1898 noting that its members were "... eminently practical rather than eloquently visionary ..."

And there was no question of the Cornish in the colony losing their individualistic spirit. A delightful story of how the Kapunda miners, almost to a man, voted *against* the candidate sponsored by the Directors in the first election after the introduction of the secret ballot, has survived. And Michael Davitt, during his Australian tour in the 1890s, visited the utopian communes that had been established along the River Murray in South Australia. While journeying from Pyap to New Residence he encountered a Cornish fisherman, who was "... a sturdy individualist, as most Cornishmen are at home and abroad..." When asked his opinion of the commune settlements, the Cornishman replied that:

Am I a "Commonist" (sic)? Not much! I work for myself, and them there "bush lawyers" up at Pyap will be all for themselves in a short time - that's sartin. It's all very well to talk and read about this Commonism, but its another thing when you come to work it out with a pick or shovel or fishing-boat. I'm no believer in these new-fangled idees, I'm not. I'm a Cornishman, I am. I have enough to do to work by my missus and myself. No sir, I'm no Commonist. Good-bye.

In the so-called "Federation Debate," too, the Cornish exhibited a fierce individualism. The Labor movement in Australia generally opposed the Federation of the six colonies -the ULP arguing that Federation would increase taxation of the workers, hit Adelaide industry through the removal of inter-colonial tariffs, dilute South Australian self-government, and allow the lucrative Broken Hill trade to be siphoned-off by New South Wales. Wallaroo was then the most cohesive working class constituency in the colony, and Federation was opposed by both the local MPs and the Peninsula branches of the AMA. However, in the 1898 referendum on the issue, the district approved overwhelmingly the Federation Bill. As in Cornwall, regional interests were considered more important than class interests - the Cornish miners believing that the inter-colonial free trade associated with Federation would boost the Peninsula's copper industry.

The Federation issue apart, however, the Cornish were identified closely with the ULP, the miners' political radicalism being evident as early as 1859 when the "Political Association" formed strong branches at Burra and Kapunda. The Assocation's political creed was payment of Parliamentary members, equal rights for all, freedom of speech and the press, an end to immigration, law reform, taxation of unimproved lands alienated from the Crown, and a belief that the "... happiness and well-being of the mass is paramount to the aggrandizement of the few." The Association's policy was to lend its support to progressive Parliamentary candidates (such as Boucaut) and, although the Association declined and disappeared during the 1860s, this was a practice which survived in the mining towns until the United Labor Party emerged in the 1890s. Boucaut's success at Burra Burra in 1868 has already been mentioned, and it is interesting to note that his successor there was the Liberal William Benjamin Rounsevell, vice-president of the Cornish Association of South Australia at its foundation.

At Kapunda, the Cornish miners acted in similar fashion, Captain John Rowe being returned as the local Member in 1862 "... largely owing to a solid miners' vote." As well as being a Cornishman (he was born at St. Agnes in 1816), he was a strong Methodist and a radical. He supported the movement for free and secular education, opposed immigration, and believed that the working miner should share fairly in the colony's mineral wealth. The Kapunda *Northern Star* noted that Rowe could hardly fail to catch the Cornishman's vote, being

... a man skilled in minerals, and mineral prospecting, well up in
his political catechism - of removal of fees and duties,
regulating of mineral leases, reimbursment or remuneration of
members ... He is a tolerably good speaker and deeply versed in
Scriptural history.

On northern Yorke Peninsula, similar criteria were important, the Cornish miners being able to exert pressure on Parliamentary candidates through their trade union. At the end of 1874 there was some talk of John Prisk standing as a union-sponsored candidate, but, as one commentator wrily observed, other union personalities such as Gill and Visick would never consent to Prisk being raised to power over their heads! The union settled down, therefore, to vetting candidates in much the same way that the Political Association had done in earlier years. In February 1875, for example, the miners declared "... in favour of Mr. Richards ..." because he supported Boucaut's policies, defended the mining interest, advocated payment of Members, and agreed with the principles of trade unionism. The fact that

105

John Richards was a Cornishman was also important - he had been born in Helston in 1843, and was a mine captain and mining journalist. He was elected as Member for Wallaroo in 1875, but resigned in 1878. Thereafter, his personal fortunes declined rapidly. He spent periods in gaol as a debtor, and in 1881 could be found "sleeping rough" in Adelaide. He died in 1913 in the Adelaide Destitute Asylum.

Another Parliamentary candidate to win the approval of the miners' union was William Henry Beaglehole: and the fact that he was born in Helston in 1834 was again an important factor in his election (he represented the area from 1881 to 1887). By 1891, however, the Amalgamated Mining Association on the Peninsula was organised and confident enough to begin sponsoring its own candidates. In May 1891 the AMA selected Richard "Dickie" Hooper to stand as an Independent Labor candidate in a by-election in the Wallaroo constituency. Hooper was born in Cornwall in 1846, was a Methodist local preacher, and a Past President of the Moonta Branch of the AMA. He won a resounding victory in the election, thus becoming the first Labor member of the House of Assembly, retaining his seat until 1902. His success was a significant event in South Australian Labor history, and heralded the growth of the United Labor Party and of a Cornish-Methodist-Radical element in the Adelaide Parliament.

In the strong working-class area of Port Adelaide there were a number of Cornishmen employed in the copper smelters and other heavy industry. The Mayor of the town in 1898 was a former mining-engineer, St. Just-born Thomas Grose, a Labor man who was "... prominently associated with Liberal and democratic associations ..." There was also David Morley Charleston, an enigmatic and rather romantic figure who talked of the need "To love, and be loved ..." and "To dream of perpetual bliss, and feel the union of souls in the one great love for Nature ..." He was imbued with "... characteristic Cornish fervour and enthusiasm ...," having been born in St. Erth in 1848. He learnt engineering at Hayle, in 1874 travelling to San Francisco and in 1884 to Australia. In 1887 he found employment in the English & Australian Copper Co.'s smelting works at Port Adelaide. There he became involved in the Labor movement and, following the introduction of payment for Members in 1890, the local Trades and Labor Council decided to adopt him as a United Labor Party candidate.

In the elections of May 1891, Charleston was elected as a Member of the Legislative Council (the upper house). His personal ideology was a fusion of liberal and socialist thought. His opinion that "To attain happiness is the end and purpose of life in high and low degree" had strong Benthamite overtones, but his views on historical development were in tune with those of Marx. He supported "... broad liberal Unionism ...," and argued that the workers should gain control of production and distribution, as a "... logical conclusion to the co-operative system ...," as a result of a slow, evolutionary process in which the unions would play a central role. He also exhibited a strong Methodist influence when he declared that "Divorce, except for proven adultery, ... I shall oppose." Nevertheless, despite this left-wing committment, Charleston quarrelled with the ULP in 1897. He resigned his seat, and refought the ensuing by-election as an Independent Liberal. His ULP opponent was trounced, and he returned to Parliament where - over the years - he drifted gradually towards the right of the political spectrum.

A perhaps less colourful figure, but nevertheless of central importance to the early ULP, was Henry Adams. He was born at Tungkillo in 1851, the son of two Cornish immigrants, Henry Adams snr., and Jane Maddern. His father was a miner in the Reedy Creek mine, and later the family moved to Moonta Mines where the young Henry was apprenticed as a pattern-maker in the Mines workshops. He was an active Methodist, and in 1881 married a Cornish girl called Ellen Eddy. He became involved in the union movement and by 1894 was President of the Trades and Labor Council. In May of that year Adams was elected to the Legislative Council, and he remained part of the backbone of the Parliamentary ULP until his defeat in March 1902.

John George Bice, born in Callington, Cornwall, in 1853, was another important Parliamentarian in this period. Although never a member of the ULP, he was a Radical Liberal and supported the Labor Party on most issues during the 1890s. His father was a mine captain, and Bice had first worked as a blacksmith at Moonta Mines. He was a Methodist, and in 1875 married Elizabeth Jane Trewenack, who also hailed from Cornwall. Bice became a local trade unionist, but he could not have been considered a dangerous militant because he was given an excellent reference by Captain Hancock when he left the mine in 1876. By 1881 Bice was in business at Port Augusta, and he was Mayor of that town in 1888-89. He first entered Parliament in May 1894 as a Member of the Legislative Council, a position he was to hold until his death in November 1923. During the 1890s he advocated a wide variety of reformist measures, always setting himself against "repressive legislation." His views on women, as articulated in 1894, were especially progressive. He declared that,

I am in favour of Adult Suffrage and because (I) believe that women are equally as intelligent, equally as capable of studying political questions, and of recording their vote as we are, I think they should have the same privileges as men in this respect. Further, without representation there is no right of taxation and under our present laws women are entrusted with the rights of property and are subjected to taxation - consequently they are entitled to rights of representation.

Acting together (but not always in harmony), Hooper, Charleston, Adams, and Bice perpetuated the tradition set by Santo and Boucaut in the early days, and established a strong Cornish influence in the South Australian

Labor movement at the Parliamentary level which survived into the 1930s and produced, in this century, the Premierships of "Honest John" Verran and R.S. Richards. Thus the Cornish miners in South Australia remained identifiably and importantly "Cornish" in their political as well as industrial activities, the development of the Cornish Radical Tradition in the colony reflecting to a very great degree the background in Cornwall. And where the development differed from that at home, it was not so much a repudiation of the Cornish experience but rather an adaption to unique Australian conditions.

PART 2

THE DIASPORA

CHAPTER 5

THE LURE OF GOLD

South Australia may have been the focal point of Cornish activity in the Australian continent in the last century, but perhaps more colourful was the Cornish involvement in the goldrushes of the 1850s - first of all in New South Wales and then, more importantly, in Victoria. Ironically, copper-rich South Australia had little gold itself to offer. But, being a colony with a strong mining and Cornish tradition, it was inevitable that its more ardent mineral-seekers would sooner or later turn their attention to gold. Some, indeed, searched the length and breadth of South Australia in their quest for the yellow metal. And, paradoxically, gold had been found in the colony some years before the first "official" discoveries in New South Wales. In 1876 it was claimed that gold deposits had been uncovered as early as 1844 but that "... the finder was not aware of the nature and importance of the discovery..." Certainly, some gold was encountered in the North Montacute copper mine, near Adelaide, in 1846, and considerable excitement was caused by a discovery on the Onkaparinga River, again near the capital, in 1849. In later years there were gold rushes to other areas in the colony, but these were all very minor occurrences when compared to the spectacular rushes elsewhere in Australia.

Consequently, those South Australians who felt the lure of gold were forced to travel to other parts of the continent. Thus it was that South Australia became intimately involved in the Victorian Gold Rush of the 1850s, with perhaps the principal South Australian contribution coming from the colony's Cornish settlers - especially the copper miners. They were amongst the first diggers on the Victorian fields and later, when their ranks had been swollen by newcomers direct from Cornwall, they formed a pool of skilled labour in Victoria from which the resurgent South Australian copper industry could draw its much-need recruits in the late 1850s and 1860s. As will be argued in later chapters, the South Australian Cornish were at a still later date in the forefront of gold discoveries in Queensland, the Northern Territory, and Western Australia - the large-scale movement of Cousin Jacks from South Australia to Victoria in the 1850s triggering a new diaspora which spread the Cornishmen throughout the continent, at the same time encouraging further migration from Cornwall to all corners of the Antipodes.

The Victorian Gold Rush commenced at the end of 1851, in response to the drain of population to New South Wales occasioned by the discovery of gold in that colony earlier in the year. Cornish miners emigrating to Australia had been urged to prospect for gold in the continent's rivers, and it was significant that the first major discovery of gold in all Australia was made by a Cousin Jack called William Tom, after he had been taught the techniques of gold-prospecting by Edward Hammond Hargraves, a veteran of the Californian Rush. Tom came from Cornish Settlement, a village founded in 1829 on the western side of the Maquarie River, 20 miles from Bathurst in New South Wales. It had been established by William Tom senior, born in Bodmin in 1791, and his friend William Lane - also from Cornwall. Tom took 640 acres of land, and was soon joined by other Cornishmen - George Hawke, the Glasson brothers, Richard Grenfell from St. Just-in-Penwith, Richard Lane, the Bray family, the Pearses, the Paulls, the Thomases, and the Oateses. Farm names such as "Pendarves" and "Tremearne" were evidence of a strong Cornish influence, and together the colonists decided upon the name of Cornish Settlement for their village (alas, it was later changed to the boring and unimaginative Byng). At its height, the village had a population of well over 1,000 people, the behaviour of the settlers exhibiting many Cornish characteristics - not least the traditional inter-Celtic rivalry with Irish, who were also represented in the locality in some numbers.

By January 1851 John Glasson, one of the Cornish Settlement colonists, was producing copper from modest workings in the area, which he had dubbed his "Cornish Mines." Other Cornishmen had been working copper and lead deposits in the district of Bathurst and Yass, and the news of William Tom's gold discovery turned their attention quickly from the base metals to the precious. By May 1851 a vigorous rush was underway, putting new towns such as Ophir, Turon, Home Rule, Hill End, Gundagai, and Gulgong onto the map of New South Wales. The *Sydney Morning Herald* declared that at Bathurst "A complete mental madness appears to have seized... the community, and ... there has been a universal rush to the diggings." This was echoed by the *Bathurst Free Press,* which wrote that "Men meet together, stare stupidly at each other, talk incoherent nonsense, and wonder what will happen next."

Amongst those who caught this gold fever and travelled to the fields were Cornishmen who had already settled in Sydney and other parts of the colony, having emigrated to New South Wales in the years before the rush. As early as 1828 advertisements calling for Cornish settlers for the colony had appeared in the *West Briton,* with many more appearing throughout the 1830s. An enthusiastic letter from one Cousin Jack in Sydney was reprinted in the newspaper in January 1838, and many Cornish set sail for New South Wales from Plymouth in ships such as the "Adromanche" in 1837, the "Eagle" in 1841, and the "Timandra" in 1842. In August 1843, after a brief respite, there was a resumption of emigration to the colony, with Isaac Latimer - the migration agent of South Australian fame - promising in February of the following year free passages for all those who were suitably qualified. A few weeks later, in early March, the "William Metcalfe" arrived in Sydney with a considerable number of Cornish on board.

In South Australia, newspapers followed closely the activities of the New South Welshmen, and at least some of the Central Colony's Cousin

Jacks were induced to travel to the goldfields. The Alpha Mine at Hill End, for example, attracted a number of men from the Burra - as always they are distinguished by their typical Cornish names; Lobb, Jeffree, Pasco, Clymo, Trevena, Hawke, Uren, Penhall, Nicholls, Roberts, Inch, Curnow, Blewett, Northey, Treglown, Bath, Carceek, Trestrail, Thomas, and so on. The departure of these miners from the Burra had a visible effect on the size and activity of the local populace, one James Jenkins writing home to Truro that "A great number have left this place to go to the Sydney gold-fields...," and noting that work was slack. The departure was mirrored in other South Australian mining towns, too. Captain Absalom Tonkin, for example, left Callington (where he had arrived, from St. Blazey, in 1847 at the age of 18) to try his luck at Turon and Ophir, later moving on to Victoria when the diggings there had become established. In all, he spent seven years prospecting in the eastern colonies, before returning to Callington to open a general store.

These Cornishmen, whether from Bathurst and Cornish Settlement, or other parts of the colony, or from South Australia, or even from Cornwall direct, fused together on the New South Wales goldfields to give the mining towns a Cornish inprint which did not altogether disappear with the principal demise of the diggings in the mid-1860s. In a version of one local folk-song, "Nine Miles From Gundagai," for example, the main character sports a Cornish name:

> As I was walking down the road
> I heard a lady say,
> "Here's Joe Rule, the bullocky bloke,
> He's bound for Gundagai."

The song is thought to date from the 1850s, the hey-day of the rush.

However, despite the initial enthusiasm for the rush, less than half the diggers on the New South Wales fields were able to make a decent living, and many found themselves working as labourers for the minority of successful men - wages being from 30s to 40s per week, with anything up to £3 for experienced Cornish miners. But still men streamed to the diggings, a flow which was not stemmed until the all-important Victorian discoveries were publicised in late 1851. Gold had been located at Clunes as early as March 1850, but little excitement was generated until the discoveries at Mount Alexander in July 1851 and at Ballarat in the following August. By the December, the influx of gold-seekers into Victoria had already begun, between 5,000 and 7,000 having arrived from the neighbouring colonies of South Australia and Van Diemen's Land. Hardships suffered by these early diggers were compounded by the drought conditions of 1852, but this had the effect of dispersing the people to new districts and thus fascilitating further gold discoveries. In the country surrounding Mount Alexander for instance, fields were opened-up at Fryer's Creek, Sailor's Gully, Campbell's Creek,

Forest Creek, Castlemaine, Ranter's Gully, and Cobbler's Gully. In February 1852 there were 30,000 people in and around Mount Alexander, and by the June there were perhaps as many as 40,000 at Bendigo. The South Australian diggers preferred the northern goldfields around the Bendigo area, and at one time there were 4,000 South Australians at Mount Alexander alone.

By the end of 1851, Adelaide newspapers were proclaiming "Gold in Abundance" in the neighbouring colony and, with business in South Australia being slack at the time, many were inclined to leave for Victoria. The Cornish miners at Burra Burra were at first cautious, sending a deputation to the goldfields to ascertain the reliability (or otherwise) of the newspaper reports. Excitement grew when the miners came back to collect their families and belongings, and the ensuing departure from the Burra precipitated a mass exodus from all the other South Australian mines as well. In December 1851 it was noted that 100 miners had arrived in Adelaide from the Burra to join ships bound for Victoria, and they were followed by others during January and February of 1852. The Burra mine soon became starved of labour. In September 1852 G.W. Vivian, the mine clerk, resigned to go to the rush, leaving as his forwarding address the "Post Office, Bendigo." A fortnight later Captain Matthew Bryant was granted two months leave so that he too might visit the diggings, and by the beginning of December the Burra was at a standstill. As the *Mining Journal* noted, "... the attractions of the gold-diggings began to tell upon the great copper mine, wooing away its lusty Cornishmen to Forest Creek and Bendigo."

There was John Treloar, for example, a Helston-born miner who in Cornwall had worked at the Trewavas Head Mine and at Great Wheal Vor before migrating to South Australia with his brother James in 1849. He was employed at the Burra as an ore-dresser, and he recalled that on the day after the news of the Victorian Rush had reached Burra he and his colleagues caught the 5 a.m. coach to Adelaide, en route to the diggings! Treloar and friends roamed across Victoria, working claims at Golden Point, Pennyweight Hill, Winter's Flat, Sulky Gully, and Buninyong, with Treloar ultimately becoming captain of the Carrick Range Mine. And there were many others like him - William Gummow from Padstow, James Bone from Morvah, John Jenkin from Perranzabuloe, John Dunstan from Stithians, William Sandow from Chacewater, Stephen Carthew from Redruth, John Phillips from Wendron, James Thomas from Ludgvan, Ralph Kestel from Portreath, all of them Burra miners - and many of them Methodists - who succumbed to the lure of gold. Those who had done well on their tribute and tutwork contracts could afford to take their families and to travel by sea, but others had to attempt the three to four week overland trek to Victoria. Many were forced by circumstances to walk, a tremendous undertaking in those early colonial days. The solitude and danger of the overland trip was conveyed in a letter written home to South Australia by one of the diggers:

113

On the 20th of last October (1851) I left Adelaide on foot, with my blanket on my back, accompanied by my dog, en route for the diggings in Victoria... After travelling the first swamp, which I forded in about two hours, knee deep; on getting out of which, all track was lost, having been grown over during the winter. Fortunately I fell in with a few natives, who put me on the right scent, and with the sun for a guide, travelled on till night, when I lit my fire and was literally alone in the desert.

For those who went in groups, and could afford to purchase a dray and bullock team, the journey was not quite so difficult. John Whitford, for example, left the Burra with his brothers James and William during 1852. They set out on 9th October, stopping first of all in the township of Macclesfield to load-up with flour and bacon. But as they travelled east they became troubled by a lack of feed and water for their bullock, and John complained bitterly about the mosquitoes. They at last found a suitable water-hole, but then soon encountered the notorious swampy country which was difficult in the extreme for the dray to negotiate. They then "... had to go through the short desert which is 18 miles ...," John noting that "... the mosquitoes and sandflies was tormenting the bullocks ..."

Further trials awaited them, but they arrived finally at the Daisy Hill diggings on Wednesday 10th November, where the Whitfords found their first gold. Six days later they moved on towards Bendigo, camping on the way at Bullock Creek where they encountered two of their Burra colleagues - Thomas Philips, who "... has done very well ...," and William Begolhall (sic), who "... has done first rate." They were told that "... almost all the Burra people are at Forest Creek; Nicholas Tambling ... William Trevena, George Roberts and all the rest of the party ..." and that "... Captain Matthew Bryant is working at Forest Creek by himself ..." The Whitfords lived frugally on potatoes and "... heavy currant cake ..." (typical Cornish fare), with mutton on Sundays. They were moderately successful, writing that "... we do not wish ourselves back to the Burra again ..." and, although they complained about the high price of flour, sugar, and tea, they concluded that "... it dont cost a great deal to live hear (sic) more than at home."

Reports such as this no doubt encouraged others to leave the Burra, and "gold-fever" spread to the other South Australian mining towns. One Cornish miner at Kapunda wrote to a friend in Truro, admitting that he "... had thoughts of going to the diggings ..." Others at Kapunda had similar thoughts, and acted upon them too - men like William Christopher from Stoke Climsland, John Rowett from St. Austell, John Rogers, John Harris, and many more. At Tungkillo the story was the same, with John Dunstan and Thomas Teague setting off together for Victoria. Henry Waters, from the Strathalbyn Mines, was on the goldfields for some seven months, as was Captain Cornelius from the Paringa. And from Callington went a number of miners - William and Josiah Odgers, Robert Peters, and the Thomas brothers from Hayle.

It was not, of course, only the professional miners who experienced the lure of gold. In Adelaide the excitement was at least as great as at the Burra, and people from all walks of life - most of whom had never heard of a gad or a whim - were tempted to try their luck as gold diggers in Victoria. At first it was only the unemployed who left, but soon it became apparent that many "honest labourers" and even the middle classes were abandoning Adelaide. Francis Treloar, a Cornishman from Penryn, wrote in his diary that "Almost all people were leaving for the diggings and Adelaide was in a state of panic." (Treloar, not surprisingly joined the rush. He sailed from Port Adelaide in March 1852, and by the end of the month had marked out a claim at Dinkey Gully). There was little chance of Adelaide becoming a ghost town, but nevertheless the effect of the rush was dramatic. In February 1852, one colonist wrote of the disturbing conditions in Adelaide:

What changes have taken place in this colony since Christmas! The discovery of gold has turned our little world upside down; thousands left the settlement for the diggings... In Adelaide windows are bricked up, and outside is written, "Gone to the Diggings." Vessels are crowded with passengers to Melbourne, and the road to the Port is like a fair - ministers, shop-keepers, policemen, masons, carpenters, clerks, councillors, labourers, farmers, doctors, lawyers, boys, and even some women, have gone either by sea or land to try their fortunes at the diggings...

In Adelaide, as in the mining towns, the Cornish were amongst the first to leave for Victoria. And although only a few were practical miners, (the mining experience of most was confined to observing white-washed engine-houses on the hill-sides at home), a fair number felt that their Cornish birth gave them a certain inate skill which would make them successful diggers on the goldfields. There was John Allen, a draper's assistant, and Robert Dunstone -a mason from Redruth - together with farmers such as Kea-born William Pedler and William Mildren from Helston. These, and many others like them, gave up their usual occupations to join the miners' throng. Indeed, the continuing departure of gold-seekers, Cornishmen and others, caused further anxiety in South Australia. The colony's copper mines were forced to suspend operations, and Adelaide was plunged into the economic doldrums. However, although little could be done to revitalise the mines, the resourceful Adelaide men were able to turn disaster into success - first of all by organising the famous Mount Alexander-Adelaide "Gold Escort" (which brought gold into the colony for processing), and then gearing South Australian agricultural production to meet the enormous demand for food from the rapidly expanding Victorian population.

Economic activity in Adelaide began to increase, and two South Australian Cornishmen who were able to exploit these new conditions to the full were Philip Santo and James Crabb Verco. Santo, from Saltash, had been Clerk of the Works at the Burra mine from 1849 until the start of the Victorian rush. He then returned to Adelaide and, anticipating the increasing demand for bread in Victoria, entered the flour trade in partnership with Verco. To ensure penetration of the goldfields marked they conducted their operations from Melbourne, and Santo used his time in Victoria to endulge in a little gold-seeking himself. James Verco, Santo's companion and business partner, was born at Callington in East Cornwall in 1815. In the 1830s he had been in Mexico and Texas (where he was embroiled in the wars of Santa Anna) but he had returned to Cornwall in 1839 to marry his childhood sweetheart, Ann Cooke of Harrowbarrow. They emigrated to South Australia in 1840, settling in Adelaide. James Verco was by trade a builder, and after the Gold Escort had become established he was commissioned by the colonial Government to construct their Adelaide smelters. He also joined Santo in Victoria for a time, on one occasion returning to Adelaide as a guard on the third Gold Escort from Mount Alexander.

South Australia's Cousin Jacks, having shown the way, were joined quickly by their fellow-countrymen who had settled in Victoria before the rush of 1851. The *West Briton* had noted as far back as May 1841 that "Several families from Newlyn and neighbourhood..." had arrived at Port Phillip (Melbourne), while five years later the same newspaper noted the emigration of one "Charles Tilly Esq. of Tremough near Penryn who takes with him the horse named Royal William, four of the handsomest pigs ever exported ... and two prime bull-dogs." And in 1849 a Mr. Allen of Penzance received from his son in Geelong (near Melbourne) a letter which-while noting wages were high - speculated that, with the growth of mining elsewhere, it was "... likely that many miners will be wanted ..." in Victoria before long. With hindsight, we know how accurate Allen's guess proved to be, and there can be little doubt that he was among the many who made their way to the diggings at the end of 1851.

Towards the end of 1852, the Australian diggers on the Victorian goldfields were joined by a new wave of immigrants from Britain, Europe, and even California. These newcomers gave the diggings a somewhat cosmopolitan flavour (soon to be heightened by an influx of Chinese), but among their ranks were many Cousin Jacks from Cornwall (and California) who served to strengthen the already sizeable Cornish community created by the South Australian diggers and earlier Cornish settlers. Like the South Australians before them, these new Cornish immigrants displayed a preference for the northern goldfields. The Cornish were responsible for the famous White Hill rush, and Long Gully - where only the Cornish had the expertise to work the deep lodes - became known as "Bendigo's Little Cornwall." Methodism was strong at Long Gully, Cornish Wrestling

116

matches were held there, carol singing parties were formed, and the miners' wives prepared traditional Cornish fare such as saffron cake.

Along with the Irish, the Cornish moulded the character of early Bendigo. And among their ranks was John (later Sir John) Quick, a principal father of Australian Federation and leading liberal statesman. He was born near St. Ives in 1852, probably at the hamlet of Trevessa, and was christened in Towednack church, emigrating with his parents two years later. They settled in Bendigo, but not long after their arrival John's father died. His mother, however, did not return to Cornwall - feeling, perhaps, that her fatherless son would have a brighter future in Victoria than at home. If so, she could not have been more correct. John Quick became involved in local politics, entering the Victorian Legislative Assembly in 1880 and soon embracing the cause of Federation. He was awarded his knighthood for his services to the cause, and he was co-author with Garran of the *Annotated Constitution of the Australian Commonwealth* (1901) - a volume which is still the basic reference for constitutional lawyers. Like many Cornishmen overseas, Quick was a confirmed liberal. Although the Cornish as a whole seemed politically less active in Victoria than in South Australia (they steered well clear of the Eureka Stockade rebellion), their liberal-nonconformist radicalism was still of importance. As late as 1886 two Parliamentary candidates standing on a Methodist-Temperence ticket in the goldfields town of Sandhurst secured 20 percent of the vote, largely through the support of the Cornish community. And John Quick, a devout Wesleyan, identified with all the great liberal issues of the day. He became a Member of the Federal House of Representatives (1901-13), and was later a Federal Judge - retiring in 1930 at the age of nearly 78, two years before his death.

But, that said, the influence of the Cornish spread far beyond just Bendigo. In October 1852 came the Oven River Rush (there were 20,000 on the River by the following January), and in the November the Korong Rush was started. New gold deposits were uncovered at Canadian Gully, Ballarat, in January 1853, and the spate of discoveries continued into 1854. The continuing success of the Victorian diggings created still further interest in Cornwall in the goldfields, and during the early 1850s Victoria replaced both South Australia and California as the prime destination for emigrant Cornish miners. Most of those who did land at Port Adelaide departed almost immediately for Melbourne, and the few who made their way to the Burra or Kapunda were greeted there with scenes of desolation. Men who had left to join the "Forty-niners" in California, such as Thomas Hosking from Gawler and Richard Snell from the Burra, came back quickly to Australia to go to the diggings. And others who had gone to California direct from Cornwall were similarly enticed. As A.C. Todd tells in his *Search For Silver*, James Skewis from Camborne made his way across America to San Francisco and the Californian goldfields. But the sight of miners leaving for Australia was too much a temptation for him, and he too set sail for the Antipodes, landing in

Sydney and then walking overland to Bendigo. He made 30 dollars a day re-working the spoils left by earlier diggers, and after a year roaming the goldfields he journeyed back to Cornwall, only soon to depart again for the United States and Mexico. And even as far away as Mexico Cousin Jacks had heard the call of Victorian gold: Captain John Dalley, a St. Austell man, and one of the first Cornish miners at the Real del Monte mines, toured the Victorian diggings before returning to his native Cornwall in August 1854.

Letters written home from the diggings helped further to popularise the Victorian goldfields. "Tell Mr. Stephens of Troon that the best thing for his sons is to send them to Australia," was the advice offered by one correspondent, whose letter found its way into the pages of the *West Briton.* A similar missive appeared in the *Royal Cornwall Gazette* in July 1852, a letter to a woman in Gulval from her sister in Port Phillip who wrote that "There are plenty of Cornish here and they are doing well. Several from the parish of Gulval have saved £10, £100 and £200 per man in a few weeks."

Considerable excitement was also created in Cornwall by Cousin Jacks landing at ports such as Penzance and Falmouth with almost fabulous amounts of gold. In April of 1852 70,000 ounces of gold, worth £25,000, was landed at the former port, where little over a year later Captain Mathews was proudly displaying his 5lb Victorian nugget. Local syndicates - such as the "Devon & Cornwall Miners Gold Company" - were formed to organise prospecting parties in Victoria, and the Cornish newspapers were full of advertisements for these enterprises. A prospectus for the "Melbourne Gold General Mining Association" appeared in the *West Briton* in February 1852, while in the April the same newspaper was able to announce not only the arrival of £80,000 worth of Victorian gold at Falmouth, but also the departure from the same harbour of the "Port Phillip Gold Mining Association's" party of Cornish miners (selected by Captain W. Richards of Redruth) in the barque "Augusta Schneider." "Wanted immediately to proceed to the Gold Diggings a few active healthy young men of good character... Apply W.N. Hosking, Falmouth," announced another advertisement, while "... a Cornishmen and two friends..." sought a party of 10 to 12 miners to accompany them to Victoria.

During 1853 a goldfields display was exhibited in Truro and Falmouth, and in February 1854 it was reported that 50 people had left St. Ives for the Australian diggings. Five months later it was noted that 100 St. Just men were at Liverpool, bound for the Victorian goldfields, and the arrival back home in the September of 20 successful Cousin Jacks caused a new flurry of excitement from Penzance to Liskeard. At first it had been mainly miners who had responded to the call of Victoria, but now Cornishmen of all classes and occupations were clammering to become Australian diggers. There was a Trewavas from Mousehole who emigrated to the Eaglehawk area of Victoria, and from neighbouring Newlyn went a party of fishermen who sailed their small boat all the way to Melbourne. There were seven of them - P.C. Mathews, Richard Nicholls, Job Kelynack, Richard and William Badcock, Lewis Lewis, and Charles Boase - who made the daring journey in 1854-55 in the "Mystery," a vessel 33ft. long and with 11ft. 6in. beam. They set-sail from Mount's Bay on the morning of November 18th 1854, and arrived finally at Hobson's Bay, Melbourne, on March 14th 1855 after a voyage of 115 days. Of the seven, five later returned to Cornwall, but Lewis died in Castlemaine and Mathews settled in Melbourne. Others tried to emulate their adventure, but the danger inherent in sailing a small boat to Australia was demonstrated clearly by the fate of the "Snowdrop," from St. Ives, which attempted the journey but was never seen again after putting-in at the Galapogos Islands for water.

There were also a number of "aristocrat diggers" on the goldfields, so that the diggers' society became classless (or at least multi-class) as well as cosmopolitan. One such "aristocrat" was William Pomeroy Carlyon, a son of Colonel Angus Carlyon of Tregrehan, near St. Austell. Apparently William was the black-sheep of the family, and Colonel Carlyon thought it would improve his son's moral fibre if he were to be sent to work in the colonies. Colonel Carlyon corresponded with Captain Richard Rodda in South Australia and John Glasson at Cornish Settlement in an attempt to find a suitable destination for his son, and he contracted finally with William Goyne and John Hammer (a miner from St. Blazey) for them to take the incorrigible William to the "... the gold diggings of Australia..." for £50. Goyne and Hammer agreed to stay with William for at least 12 months after their arrival in Victoria. By early 1853 they were at the Ballarat diggings. William wrote home to his father complaining that his hands were "... getting very stiff with work..." and expressing regret for the wild life he had led at home. Goyne and Hammer promised to stick with William "... through the rough and smooth..." but the young Carlyon caught the dreaded "colonial fever" (typhoid) and was dead by the July. His death was made all the more ironic by the fact that he, Goyne, and Hammer had between them amassed £1,333 worth of gold-dust in their short time in Australia.

For those who journeyed to the diggings by sea, whether from Port Adelaide or Plymouth, their first experience of Victoria was Melbourne - which in the gold rush days was, as Geoffrey Blainey has said, almost a second San Francisco. Cousin Jacks from Truro and Burra, Camborne and Kapunda, Bodmin and Bathurst, congregated in the great city, many of them no doubt staying in the "Great Western Hotel," an establishment run by two Cornishmen and which advertised in the *West Briton*. Peter Pascoe, a tailor from Helston, was in Melbourne during 1852. He wrote that "Pecunary matters is good," and added that most people were "... Rich in a manner of speaking and as independent as Lords..." Already he had come across two friends from Helston, John Ellis and Edward Toy, and he estimated that "... there is upwards of a Hundred with their families from Helston in Melbourne." The *West Briton* also noted Pascoe's success, explaining to its readers how he had thrown a "sumptuous dinner" for 20 of his friends from
119

Cornwall.

On the diggings themselves the Cousin Jacks, whether from South Australia or Cornwall direct, tended to "stick together" - not only at Long Gully and Bendigo, but in most places that they ventured to, a fact epitomised in the comments of one Peter Matthews who wrote from Ballarat in 1853 that "The place in which we live is called Cornish Town, on account of the inhabitants being nearly all Cornish." Similarly, Captain George Read from St. Day wrote home to say that he and his friend Mr. Martin (a former Sithney inn-keeper) were part of a group of 30 Cornishmen, 12 of whom were from Gwennap and Stithians, and adding that another party of 84 West Cornishmen had just arrived.

Stephen Curnow, who came from Bog Farm near Marazion and had emigrated in Brunel's "Great Britain" in 1852, wrote that he "... fell in with (a) Cousin from Gulval...," using the term "Cousin" to signify "Cornishman" (c.f. Cousin Jack). And from Friar's Creek in January 1853 he wrote tellingly that "Cousin William Roach and his Mrs visited last Sunday has (sic) it was Ludgvan feast." By the August he had encountered other "Cousins" - "John Lawry, Tom Polmear... Joseph Williams... (and) some new arrivals they where (sic) from St. Ives, Lelant and St. Hilary." Curnow moved on to Spring Gully with Joseph and William Williams, two Cornishmen, and in September 1854 was joined by Thomas, John and William Roach who had just returned from the Burra. He made friends with "... many others from Ludgvan too numerous to mention..." and in February 1855 noted that "Cousins John and Thomas Roach left here a few days after Christmas for the Burra I think the(y) intended comming (sic) again as soon as winter sets in." Stephen Curnow returned to Cornwall during 1857, but his colleague John Williams was still on the goldfields as late as March 1860. He wrote to Curnow from Reef Pleasant Creek, remarking that "... thear is not manny cornish men cas it is a out of the way place..." but adding that "... Phillip Williams is still in Ballaratt keeping a otell I think he is getting on very well..."

Henry Giles arrived in Victoria in 1854 with 30 other Cornishmen, "... principally from St. Day and Chacewater and three from St. Ives - Thomas Bennett, Mathew Thomas and William Daniel." Amongst the 15 or so Cousin Jacks he accompanied to Creswick were "... Richard Eddy from Treen and Matthew Thomas from Treen and David Eddy from Bosigran... Matthew White, Richard Eddy from Bosigran, John Hosking from Treveal... and Arthur Chellew from Zennor Church Town" - all the settlements mentioned being located on the five mile stretch of road between Morvah and St. Ives in far-western Cornwall. Giles added that there were also "... two miners, one is from Sancreed called George Thomas, the other from St. Day called Richard Harvey...," and explained that

George Thomas is my Mate, he and me do belong on one pit we

120

expect to Bottom next week which will be about 60 feet. John and Richard Harvey is together on another pit. Arthur Chellew is with William Wearne from St. Just in another pit, so we shall all share alike on the gain.

In May 1855 Giles and his companions, together with "... a young man from Penryn...," had gone to try their luck at Daisy Hill. Giles, however, had injured his knee and while confined to his tent to rest contracted dysentery, from which he died on the morning of May 16th. A friend wrote to Giles' parents in Cornwall, trying to console them by saying that their son "... had a remarkable easy death, died without a struggle or a groan..." and assuring them that "... the neighbouring people were very kind to him during his illness, there being many Cornish people near, especially Mary Ann Simmons from Madron Church Town..."

Henry Giles' death, like that of William Pomeroy Carlyon, bore witness to the fact that life on the diggings was never easy. Typhoid and dysentery were constant perils, while violence and drunkeness aggravated the already unpleasant social conditions. The mining camps which sprang up around the diggings were by their very nature temporary affairs, breeding grounds for both disease and vice. One South Australian digger wrote home in 1852, describing the appearance of Forest Creek:

(It is)... a large crooked gully, almost a valley, the hills side by side not so much in ranges as in isolated mountains. Cross gullies running into it at intervals. A road-way along the creek with stores by the side of it, built of wood, bark, iron, or canvas - all with flags of different colonies flying. Tents of all different shapes, sizes and materials here and there, some by the road like a street, and some scattered about the gullies and hills in every direction.

The diggers who inhabited these make-shift towns of calico and timber were,

...as dirty with clay and mud as they can be, as thick as ants, and as busy as bees - such they are on week-days. But on Sundays - red shirts, blue shirts, red comforters, red night caps, and indeed every colour of the rainbow of some, while others dress more soberly; but dress as plainly as they may, they look strange and rather wild, as not one in 20 shaves.

And to survive, the diggers had to be tough. As another South Australian wrote:

We had to fight for our claim. It was nothing uncommon to find a big rough fellow working your hole and disputing your right to it. This, in the early days, had to be decided by a stand up fight, and I must say that it was a quick, if not just, way of arriving at a decision. The diggers would always see fair play.

The violence and rough justice of the goldfields was a theme returned to again and again by contemporary observers. As early as December 1851 the *Register* had warned South Australians that "... life at the diggings must be a life of perfect wretchedness...," there being a need for "... incessant watchfulness with ... frequent resort to strong if not violent measures in self-defence." The *West Briton* informed gold-crazed Cornishmen in December 1852 that one John Semmens, from Ludgvan Lease, had been murdered in Victoria for the 60 ounces of gold dust that he had won, and less than three months later the same paper noted the very similar murder of Henry James, from Gear in Gulval parish, by his "friends" on the diggings. Stephen Treseder, formerly of Truro, wrote to the *West Briton* from Victoria declaring that "The curse of the country is drink," and a commentary on the visit of a deputation of Wesleyan ministers to the goldfields had this to say on the conditions they encountered:

Parties of eager gold-seekers were passed on the way, generally looking most wretched and forlorn. Forest-creek and Bendigo diggings were successively visited and examined; and many old friends, especially from Cornwall, gave the Deputation a cordial and hospitable welcome. The social state of the gold-digging community is described in no very inviting terms. Many and terrible are the hardships to be undergone... and ... that eloquent peacemaker, the revolver, ... has found a lodgement in every tent; Drunkeness appears to have latterly become common, and nearly all the breaches of social order and morality are ... traced to its accursed influence.

One might expect the Methodists to have taken the dimmest view of life on the diggings (and indeed the whole ethos of goldseeking), but that their lamentations were not exaggerated is more than confirmed by the series of entries in the diary of Francis Treloar, one of the many South Australian Cousin Jacks in Victoria.

6th June 1852 Last night a man was shot while stealing a bag of flour...
16th July 1852 Man killed in a hole at Wattle Flat.
12th August 1852 Mr. Manchester, a storekeeper drowned while crossing Sawpit Gully, 8 miles from diggings.
15th August 1852 A man near us was shot.
17th August 1852 A women fell in a water hole and was drowned. Supposed to be under drink.

122

9th November 1852 Man killed today by cart upsetting on him.
11th November 1852 Man killed going down a hole. Windless fell on him.
12th November 1852 Another man killed by hole falling in on him while working.

Francis Treloar himself survived the rigours of the goldfields, and by January 1853 was working as a teamster at the Burra Burra mine, back in his home colony. Other Cousin Jacks were less fortunate, one article in the *Mining Journal* describing in somewhat excessive language the demise of a young Cornish miner who, defying the protestations of his parents and sweetheart, had made his way to the Victorian goldfields:

He repaired at once to the diggings, where, as he had expected, his profession stood him in good stead. Strange to say he was singularly fortunate, but so much the worst for him. How often is apparent good but a deceptive reality! How frequently, too, we mistake the shadow for the substance! His success excited him to over-exertion and anxiety to secure his glittering prize. These fostered the worm already praying on his vitals, but he heeded it not - gold, gold, gold was ... the only thing in the world that had charms for him... Comforts he cared not for, sympathies he sought not, health he disregarded; all these would be on his return. Vain hope, deceptive vision! It was too late.

Food shortages, and the attendent high prices, were a source of further misery - one observer in 1854 noting that a "... feed of corn ..." was between 5s and 7s 6d, while tea was 4s, a bottle of ale 6s, and stabling for a horse for the night 20s! Such prices, it was said, rendered "... a profit of at least 1,000 percent, and ... were moderate compared with charges a few months before!" Profiteering there certainly was, and before his death Henry Giles had railed against the malpractices and evils of many of the diggers. He wrote that

This is a terrible place for sin and wickedness... there is thousands here who don't pay any attention to Sundays no more than another day. They go shouting and cutting wood and spending the Sabbath in a most fearful way. Butchers do kill on Sundays the same as week days... There is no respect of persons here, Jack is as good as is (sic) Master. There is no use for a man to come here unless he is steady and not given to drink. There is hundreds here who spend their money as soon as they get it....

But as at Burra Burra and in Cornwall, some of the worst excesses were eradicated, or at least controlled, by the work of the various Methodist denominations. Make-shift chapels were erected on the actual diggings - George Prout, a young lad from Camborne, was one of the first to attend the Wesleyan chapel at Forest Creek, and Peter Matthews wrote that at Cornish Town (near Ballarat) "There is a large Tent erected for a place of Worship not far from us. Its dimensions are 46 feet in length and 36 in breadth." There were a few ordained Ministers on the goldfields, but the great bulk of missionary work was performed by ordinary lay preachers who arose from the ranks of the diggers themselves. When James Skewis arrived at Bendigo from California he was delighted to find three "splendid good preachers" of the names of Stephens, Rowe, and Jeffery. This latter was probably James Jeffery, a Cornishman who had emigrated from Illogan to South Australia and had worked in the Burra mine. He joined the Victorian Gold Rush, not returning to the Central Colony until 1872. He died finally at Moonta in 1877, aged 61. Blamires and Smith, in their *Early Story of The Wesleyan Methodist Church in Victoria* of 1886, described Jeffery as the typical Cousin Jack:

He was a short sturdy man with dark hair and features, twinkling eyes, and a pulpit and platform manner that was quiet and modest. His homely talk, quaint repartee, Cornish brogue, unexpected turns of speech and pertinent illustrations, conjoined with him a great power with the people in the mining districts.

James Jeffery held the first-ever religious service on the Bendigo goldfields and his popularity and success no doubt stemmed from his ability to communicate with the diggers (he was, after all, one of them) through the medium of the contemporary parable or allegory. He would, for example, declare that:

You diggers mark out a claim and put down your pegs near to a mount, say that it is Mount Alexander, or Mount Tarrengower, or the Wombat Hill, and you will go to work in the hope of finding the gold, and some of you come on a rich patch, and others sink "schicer" holes; 'tis terribly uncertain about finding the gold; but I'll lay you on to the best place. Here, you diggers, come work out a claim by Mount Calvary.

Another Cornish preacher was William Moyle, who also journeyed to the diggings from South Australia. Although he earned a reputation as a man of God, his gold-seeking was dogged with ill-luck, it being said that "Good father Moyle was always richer in grace than gold..." He went first to Bendigo, moving on from there to Ballarat in 1853, and soon returning in disgust to South Australia. Yet another lay preacher was John Trevellyan -he, however, was dubbed by some "... a religious charlatan..." and considered by many "... an unreasoning Cornish miner." John Sherer, in his *The Gold-Finder of Australia* of 1853, had several amusing stories to tell about Trevellyan. He first encountered the bogus preacher at Forest Creek during a funeral ceremony. Trevellyan had agreed to hold the service "... for a small consideration...," and during his address launched out at the wantoness of the diggers, exclaiming that "... gold is the murderer. He is the Appolyon that is burying your bodies in the earth and plunging your souls into hell." But while in the midst of this tirade, Trevellyan was recognised by a fellow Cornishman who knew him to be an unscrupulous and disreputable fellow. This Cousin Jack (Simon was his name) called out to Trevellyan: "How be the goats and kids at Stithians?" Sherer recorded Trevellyan's reaction, and his subsequent discussion with Simon:

"O Simon, Simon!" replied Jahn (sic), with well-feigned astonishment, and with a view to preserve the character in which he had appeared, "be ye here, too, amongst the worshippers of Mammon?"
"Deed am I," said Simon, "and how be ye here, if I should be so bould?"
"I came," said John, "to call sinners to repentance."
"They you've come to a fruitful spot," returned Simon, "for there's plenty of 'em here."

Simon later confided to Sherer that Trevellyan was a "back-slider" (to use the Methodist terminology) who had at one time been converted, but had since fallen into evil ways. He had had his name read out in chapel three times for drunkeness, twice for swindling, and at least a dozen times for forging letters and "... evil speaking..." But for all that, Simon admitted that Trevellyan was "... a good preacher ... and has been a great means of saving souls in Carnwell (sic)." Even Sherer was forced to note Trevellyan's moral strength, writing that he

... had the merit of courage, in placing himself (and preaching) in Friar's Creek, where ... the state of society was low in the extreme, and where the greatest insecurity of life and property existed. Bands of the blackest ruffians under the sun were well known to be continually haunting this spot, who every night, and sometimes even by day, committed the most impudent robberies.

Thus the Cornish preachers, whether bona fide such as Jeffery and Moyle, or bogus like Trevellyan, performed a vital social function in ministering to the diggers. But poor social conditions derived not only from the violence, drink, and other vices, but also from the plain fact that most of the diggers were unsuccessful in their search for gold. It was certainly true, of course, that Cousin Jacks - whether from Australia, California, or Cornwall direct - were in the forefront of the gold discoverers of Victoria: Alfred Chenery was the first to prospect successfully on the Upper Goulburn River in 1853, William Jewell and John Thomas discovered Fiery Creek in August 1854, Jack Bastian started Sailor's Gully, John Northey was responsible for the Stockyard Creek Rush, and so on - with names such as Richard Higgs, H.C.P. Pollard, John Roach, Thomas Kemp, W. Pawley and W. Polkinghorne being prominent amongst the list of other official gold discoverers.

There were, too, cases of startling personal success. As the *Cornish Telegraph* noted in January 1872, a good example was that of the Rowe bothers, who worked Ferron's Reef near Castlemaine for all of 13 years, at an incredible profit of £400 per week. Similarly, both the world-famous "Welcome Nugget" and "Welcome Stranger" finds were made by parties of Cornishmen - in 1858 and 1869 respectively. The *West Briton* got news of the former from a certain Captain John Ivey of Camborne who had received a letter from his son in Victoria explaining that a giant nugget, weighing some 185lbs, had been found at Ballarat by a party of 22 Cornishmen - nine of whom were from Illogan, including William Jeffery (an early Californian digger) and his brother Richard. The "Welcome Stranger" was an even more spectacular find, at well-over 200lbs being easily the biggest single nugget of gold ever discovered in Australia (or indeed the world). Its lucky finders were two Cornish miners, John Deason and Richard Oates, who came across the nugget only a few inches below the surface at Bulldog Gully, one of the central Victorian goldfields. Fortunately, a professional photographer by the name of Parker was on hand to record the event, and one photograph depicting Deason and Oates, their families and friends, and of course the nugget, has survived - giving the modern historian a very human insight into the events of that day in 1869. Deason and Oates broke pieces off the nugget (not too large, one supects!) to give to their friends before dividing it into three parts to transport to the bank at nearby Dunolly, where they sold it for £9,500. When smelted down, the "Welcome Stranger" yielded 2, 268 ounces of pure gold!

There were, of course, other success stories. William Mitchell, from St. Agnes, made his fortune as a carrier on the goldfields and went into farming, his "Trevellas" property being one of the largest in Victoria. William Prout, a digger from Camborne, was similarly successful, eventually taking up land along the Bet Bet Creek, and John "Cranky Jan" White from Trewellard (near St. Just) made enough money at the diggings in 1851 to bring out his whole family from Cornwall and retire in comfort. A Cornish party from Willunga (where there were several Cornish-run slate quarries, among them the "Delabole") in South Australia was also particularly lucky, an article in

A Cornishman, William Chenoweth, writes to his brother in the neighbourhood of Camelford, from the gold diggings. He says "I am happy to inform you that I have been successful at the gold diggings. I left here (Willunga) with Daniel Oliver, F. Martin, Robert Sleep, and Richard Polkinhorn (from West Cornwall). We were wanting from here about nineteen weeks. We were ten weeks and four days at the diggings. We worked very hard for the first seven weeks, and made about 1½ oz. of gold each; the other three weeks and four days we made about £4,500, making £900 each. We are truly thankful for our success, for although we have done this, we are one party out of a thousand to do so well. We landed at Adelaide this day week, and I deposited in the bank £853; in about two months I intend going again."

As Chenoweth admitted, however, he and his colleagues were amongst the lucky few. It was estimated at the time, for example, that for every 40 South Australians at Mount Alexander, one was making from £100 to £200 per week, seven making double or triple normal wages, seven fair wages, and the rest very little more than subsistence or even nothing at all. A number had wealth within their grasp but lost it through recklessness or sheer bad luck. Such was the case of Thomas Major, who was born in Cornwall in 1834 and arrived in Australia with his parents at the age of 13. From 1847 until 1852 he worked in the Burra Burra mine, but then made his way to the Victorian diggings. After four months at Friar's Creek he returned to the Burra with an amazing 18lb of gold. Flushed by his success, he returned to Victoria with his brothers, and not long after opened the first brewery at Ballarat (not all Cornish opposed "the demon drink"). Selling this, he next invested his capital in an ambitious development of the North Grenville gold mine (Cornish-named, be it noted). But the resounding failure of that venture left Major almost ruined, and it was only his own enterprise and natural business ability which enabled him to recuperate at least some of his losses through the operation of a wood-carting service and the later aquisition of a livery stable-cum-corn store in the goldfields town of Clunes. Even so, he was only marginally successful and worked for some years as a common miner in New Zealand, Victoria and Moonta Mines before again achieving commercial prominence, becoming finally Mayor of Port Pirie (in South Australia).

It appears that the most successful diggers were those who were the first on the goldfields, those who had the "pickings" of the most accessible and easily-won deposits. Nevertheless, even if one South Australian could write in April 1852 that Mount Alexander was "... the most wonderful place in the world...," other early diggers recorded that "... hundreds are there who curse the day they left South Australia..." And in August 1853 the *Register* drew what it considered a telling comparison between the fortunes of the Cornish miners at the Burra and those of their compatriots in Victoria:

Copper Digging v Gold Digging - a father and three sons, tributers at the Burra Burra Mines, earned during the last "Take" - eight weeks - £24 5s 4d per week; and another party of men made nearly £4 per week during the same period - both parties being employed in working "fetches" in the shallow levels. Such wages as those, with comfortable houses and other advantages existing at Kooringa, will bear favourable comparison with the vicissitudes of the goldfields in Victoria.

Peter Matthews, writing home to Cornwall from Ballarat in July 1853, noted that "Gold digging his (sic) Chance work, a person might make a fortune in the first pit he sinks or he might sink 20 and Get nothing," while in the following year Stephen Curnow wrote that many were "... wishing themselves Home if the(y) had been Transported for Life the(y) could not Look more sad. Great numbers of Cornishmen lose all confidence as soon has (sic) the(y) loose (sic) sight of Engine houses and white jackets..." Henry Giles, in February 1855, recognised that the easy days of gold-seeking were gone for good, and in a letter to his parents in Cornwall emphasised that,

I would not advise anyone at home to think that they are sure of making their fortunes by coming to Australia. There is not the chance now that there was three or four years since. A man could not miss in them days. In fact the Diggins are still very rich but very likely there is fifty to one on the Diggins now to what there was four years since. However there is still a better chance for a careful industrious man here as what there is at home.

As time went on, the chance of "striking it rich" on the goldfields became more and more remote, with Richard Hancock writing home to St. Austell from Bell's Reef (in the Tarrengower goldfields) in 1862 saying that he had made only "... just enough to pay for my meat..." and that now, instead of digging on his own account, he was working on "tripped" (i.e. tribute) for someone else - a situation he seemed to prefer, because it was like being at home. However, he could not have been entirely satisfied with his performance in Victoria, for by July 1864 he was in Adelaide (he was employed there in the Glen Osmond quarries, where sadly he was killed in an accident in the following September). Increasingly, there were fewer diggers in Victoria undertaking their own prospecting work, and those who continued to do so had to endure - as one contributer to the *Wallaroo Times* noted in February 1867 - conditions that "... Cornish miners at home would call slavish..." Many diggers, especially the Cornish, resented the loss of individual freedom involved in having to work for a syndicate or company - and as a kind of compromise many gold mines introduced the Cornish tribute system of employment, where the miner retained a degree of independence. This also facilitated the development of deep, quartz mining in Victoria during the 1870s, as it prevented a wholesale exodus of skilled Cornish miners to South Australia.

Indeed, after 1855 - and especially during the 1860s and 1870s - Victorian gold-mining became an organised and highly skilled industry, a period during which the Cornish hard-rock miners came into their own. Geoffrey Blainey has written of "... the bold Cornishmen and Californians, the innovaters in Victorian mining," and noted that "... Bendigo's great Cornish managers and miners mastered easily the problems of deep sinking." The Victorian Parliament, in an effort to promote gold-mining, passed legislation making it possible for companies to be formed under the Cornish cost-book system, and Cornish diggers themselves began to organise into syndicates, using their savings to buy the necessary equipment for deeper mining. Peter Matthews, at Ballarat, joined a party of 22 other diggers in 1855 (including John James, a mason from Polgooth) and together they purchased a steam pump engine. Matthews observed that the mines were becoming "... deep and unpleasant..." and added that "There is a great number of Horse Whims and Steam Engines now in operation at Ballarat." Some of these engines were obtained from Cornish-owned foundries in Victoria, such as Abraham Roberts and Mitchell & Osborne, while others were imported direct from Cornwall. In 1860, for example, Holman's Tregaseal Foundry at St. Just-in-Penwith constructed at 17 inch horizontal engine for the Oven River diggings in Victoria.

This new period in Victorian gold-mining was highly successful. Open-cut methods were introduced at Ballarat and Bendigo from about 1880, and hydraulic procedures for dredging and water sluicing followed a decade of two later, these two towns leading Australia in gold-mining technical development for more than 30 years. They also developed as major centres of population and, unlike most mining areas, were rather elegant towns and became well-known for their "culture and learning." Ralph Stokes, a former editor of the Johannesburg *Rand Daily Mail,* in 1901 described Bendigo as "... one of the most beautiful mining centres in the World," the town certainly contrasting strongly with the grim austerity of, say, Camborne, Moonta, or the Rhondda. But deep mining was never easy, as evidenced by Australia's first major gold-mining disaster in December 1882 when water from old workings broke through into a new shaft at Creswick, drowning 22 miners. Indeed, Bendigo's gold mines became amongst the deepest in the world. At the turn of the century, for example, the Victoria Deep Quartz mine had reached a depth of 1.28 kilometres, while the Lazarus New Chum was down

1.14 kilometers. They were also very rich, the Port Phillip and Colonial Gold Mining Company's mine at Clunes yielding 14 tons of pure gold between 1857 and 1886, worth £2,049, 415.

At the same time as this consolidation and re-organisation of the gold-mining industry in Victoria was occurring, the workers involved in the industry were also consolidating and organising. Although the tribute system preserved elements of their independence, many miners were now working as employees of the larger companies rather than operating, as they had done hitherto, as self-employed diggers. Inevitably, therefore, a distinction between "capital" and "labour" became apparent, this finding expression in the emergence of trade unions on the goldfields. And at the same time as the Moonta and Wallaroo Cornishmen, in South Australia, were taking their first faltering steps towards the creation of a vibrant and cohesive union movement, so their colleagues in neighbouring Victoria were setting about the foundation of their own Miners' Association. As noted above, the Cornish-nonconformist political tradition made an impact (though somewhat limited in nature) on Victorian politics, and this same radicalism was also injected into the union movement. Unlike South Australia, the miners' unions in Victoria were not identified exclusively with the Cousin Jacks, but Cornishmen - and their influence - can be detected, especially in the early days.

In the summer of 1871-72, the "Bendigo Miners' Association" was formed - the first union of its kind in Victoria. It was immediately very successful, attracting a great many local miners into its ranks, and engaging in fruitful negotiations with the mine managers. Its most spectacular victory was the achievement of the eight-hour day for its members, but it also influenced profoundly the Regulation of Mines Act (made law by the Victorian Parliament in 1874), while also ensuring the Association's continuing popularity in the mining communities by campaigning successfully for the blocking of Chinese labour (on every Australian field, the Chinese - who were prepared to work longer hours, in worse conditions, for less money - were detested by the Cornish and other British-Australian miners).

The success of the "Bendigo Miners' Association" led, in 1874, to the formation of the "Amalgamated Miners' Association" (the AMA), which shortly became the leading union for miners throughout the continent. The heart of the AMA was the Victorian goldfields town of Creswick, where the union had its headquarters, and the first President of the Creswick Branch of the AMA was a Cornish miner called John Sampson. Sampson's brand of unionism was liberal, moderate, and chapel-oriented (again reminiscent of South Australia), and it is interesting to reflect that Sampson's grandson was none other than Sir Robert Menzies, perhaps Australia's greatest-ever Prime Minister. Though Menzies was noticably to the "Right" of his grandfather and that generation of Cornish-Australian miners, he shared both their "small c" conservatism and their liberal outlook, producing that subtle blend of political pragmatism and progressivism for which the Cornish were known. So is it too much a flight of fancy to point to his grandfather and his quarter-Cornish background, when seeking at least some clues to the origins of Menzies' attitudes and characteristics? (when Sir Robert died in 1978 one Cornish newspaper recalled in its obituary article that he was a "...grandson of Cornwall").

During the 1870s, at the dawn of this new era in Victorian mining, immigration into South Australia was severely reduced, and there was consequently a considerable influx of miners into Victoria - many of them going to the Sandhurst mines, Bendigo, where George Lansell, a local magnate, actively promoted migration from Cornwall. Several mine names, such as the "Cornwall" and the "Cornish United," were evidence of a continuing Cornish influence, and on one occasion it was reported that 73 men from St. Just alone had been counted during an evening at a Bendigo hall. As late as 1886, it was noted in the *Cornish Telegraph* that "A recent emigrant to Australia from Boscaswell writes home to say that he made a call on one his neighbours ... Jammie (James) Lanyon. He there met with fifty Boscaswell men in one room ..." The records and minutes of gold-town institutions also attest to the presence of many Cousin Jacks in Victoria long after the early Rush days. The quarterly meeting reports of the Creswick Primitive Methodist Church in 1885-1888, for instance, are full of obvious and typical Cornish names - Grenfell, Jenkin, Rowe, Moyle, Rodda, Coad, Champion, Berriman, Tregloan, Langdon, Hawke, and so on. And in addition to the flow of immigrants direct from Cornwall in the 1870s, there were, too, a number of smaller rushes which revived the attractions of the diggings from time to time, thus encouraging a continuing (if somewhat insignificant) movement of Cornishmen from other parts of Australia - particularly South Australia - to Victoria. The Snowy River Rush, in the Victoria-New South Wales border country in 1860, was the most important of these. A large number of miners left Burra and Kapunda for Snowy River during April 1860, the *Northern Star* recording the unfortunate story of "Poor Bill Tremaine" who, while away at the Snowy River diggings, forgot to keep up the payments for his land at Kapunda, and on his return found that it had been sold in his absence!

However, despite this enduring Cornish influence in Victoria (and it would be a mistake to underestimate its extent), it is also clear that a great many Cousin Jacks who had arrived in the colony during the early Rush years were later attracted to South Australia as the copper industry began to revive. By the mid-1850s, diggers were beginning to realise that it was becoming increasingly difficult to be successful on the goldfields, and were therefore easily enticed to the Burra, Kapunda and other South Australian mines by promises of high wages and the prospect of a more secure form of existence. And, moreover, the opening-up of the northern Yorke Peninsula mines during the early 1860s provided a new focal point for the Cornish in Australia,

131

so that Moonta and Wallaroo replaced Bendigo and Ballarat as the principal Cornish magnets. As early as December 1852, Henry Ayers - the South Australian Mining Association Secretary - had hoped that, with there being "... a reaction in favour of Adelaide just now...," the Burra miners would come back to their home colony. But he was mistaken, and in the following June he wrote to Captain Matthew Bryant (then still at Forest Creek) asking him to try to recruit for SAMA some of its former employees. Ayers had been led to believe that,

... the yield of gold has considerably diminished and with it the earnings of the diggers must be much less... I am anxious to know from you whether from your experience you consider this to be the true state of the case or not and whether you think that a number say fifty or a hundred of our old miners could be induced to return to the Burra Burra if we guaranteed them (£3) three pounds per week for say the first six months after their return. You know the sort of men we most require in recommencing operations; they should be good active men.

Although Bryant took up his task with enthusiasm (he wrote twice to Ayers during the July), Ayers found that a number of the men were returning on their own account, without need of financial inducement, and so he instructed Bryant to postpone recruitment for "... the next three or four months ..." to see whether the men who had come back would stay. As it happened, too few miners returned to all low work in the deeper sections to be restarted, and so soon Ayers was writing again to J.B. Wilcocks in Plymouth and Captain Bryant in Victoria, asking both men for Cornish miners. Work recomenced in earnest at both the Burra and Kapunda mines in early 1855, with Henry Ayers writing to Captain Bryant on 2nd January explaining that Cornish miners would now be engaged at rates of between £8 and £10 per month, and adding that "We shall commence forking the water tomorrow or next day." News that the Burra was back in full operation attracted many Cousin Jacks from the diggings, and by the end of 1855 most of the old Burra hands - together with others of Victoria's Cornish miners - were hard at work in the Central Colony's copper mines.

One of the major problems faced by the Moonta, Wallaroo and other northern Yorke Peninsula mining companies during the early 1860s was a shortage of manpower. They reacted by "poaching" miners from Burra, Kapunda, Callington and elsewhere, and by encouraging immigration from Cornwall, but at the same time they recruited extensively in the Victorian goldfields. As early as April 1861 the *Register* had noted that "The news of the Wallaroo Mines having reached the diggings in Victoria, we find that several Cornishmen have left their gold-diggings to come here and work at their old occupation (copper mining)." And by 1864 both the Moonta and Wallaroo companies were employing agents on the goldfield to engage Cornish miners. The Wallaroo Proprieters placed an advertisement in the Victorian newspapers which read "Miners wanted for the Copper mines at Wallaroo. Passage Paid. Apply to Messrs. Wooley and Nephew, Melbourne," and the Moonta Directors sent their agent, S.R. Wakefield, to tour the goldfields. He was instructed to engage "... some 200 to 300 miners ..." and to persuade others to come to the Peninsula "... on their own account." The recruits were to be "... from the Mines of Cornwall and Wales, and it is only to be experienced and respectable miners..."

Wakefield reported from the goldfields in mid-August that he was experiencing some "... difficulty in obtaining the number of Cornish Miners required...," but the Board insisted that he seek only "... Copper Miners and not engage Gold diggers" (i.e. they required professional as opposed to amateur miners). They also felt that in many areas of Victoria there were "... Cornish Miners ... indifferently employed who would be only too willing to travel to Moonta." Clearly, their confidence was justified for by 30th September 150 miners, together with their families, had already arrived at Moonta from Victoria, more than 70 of these having travelled directly to Port Wallaroo on the steamship "Coorong."

The next major movement of Cornish miners from Victoria to northern Yorke Peninsula occurred in 1868, when in May Captain Datson was sent to recruit on the goldfields. By 22nd June "... 35 Cornish Miners..." had set sail from Melbourne on the "Coorong," men who were considered "... smart looking fellows ..." by the company secretary when he inspected them. In late July, 50 more miners arrived on the "Coorong." They were followed less than a month later by a further contingent on board the "Kangaroo," and still more arrived in late August on the "Aldinga." Another 21 adults and one child were said to be en route to the Peninsula by sea, and a further party was travelling overland from Melbourne. After 1868, Cornish miners continued to journey from Victoria to the Peninsula, but on their own initiative. For example, a party arrived from the diggings in November 1872, and another group arrived from Ballarat in February 1873. Thereafter, with the growing demand for Cornish hard-rock men to work the deep quartz leads in Victoria, this movement subsided, although as late as February 1881 Captain Hancock was on the goldfields recruiting miners for the Wallaroo Mines.

The attempts of the Burra, Moonta and Wallaroo mines to recruit Cousin Jacks from Victoria certainly helped to redress the "balance" between South Australia and her neighbour, and to re-establish the Central Colony as the principal home of the Cornish miner in the Australian continent. But it did nothing to dim the lure of gold, and thus the efforts of the copper companies were matched by equally ardent attempts to discover and develop goldfields in South Australia itself. Indeed, the Victorian Gold Rush had been, paradoxically, an impetus to gold prospecting in the Central Colony. After numerous false alarms, the first real discovery of payable gold was made at last in August 1852, at Echunga in the Adelaide Hills. The modest rush attracted some Cornishmen, as did the very similar discoveries in the Hills at Forest Range in 1854 and at Jupiter Creek in 1868. The latter find attracted many miners from northern Yorke Peninsula, and other Cornishmen made their way to the Barossa goldfields - north of Adelaide - which had been started in the same year.

Further north, in the Ulooloo fields near Burra Burra, and even further north still - at Waukaringa, Mannahill, and Tarcoola - Cousin Jacks could also be found in their search for gold. But despite this activity, very few South Australian Cornishmen ever made a living - let alone a fortune - from diggings in their home colony. South Australia, and in particular its Cornish community, had always to look far beyond its borders to find fields of sufficient wealth to satisfy the "lure of gold" it experienced. This had led the Cousin Jacks to trail-blaze across Victoria, and it would later send them wandering the arid wastes of Western Australia and the Northern Territory, and even into the tropical climes of Queensland.

CHAPTER 6

THE BARRIER AND THE HILL

The railway line from Adelaide to Broken Hill follows, north from the Burra, the old diggers' track - a trail first trod in the 1860s, but not well used until two decades later. Almost the entire route lies within South Australia, that vast arid tract of 100 million hectares which borders every mainland State and Territory. Only the last 20 miles or so of the line are within New South Wales, the railway crossing the border at Cockburn Siding. The journey is a long one, taking almost all day, the Adelaide train having stopped at seemingly countless wayside stations, and there being a lengthy period on the windswept platforms of inhospitable Peterborough while awaiting the late-running Transcontinental connection. Invariably the train arrives in Broken Hill at the dead of night, only the head-frames - illuminated by the arc-lights of the mines - hinting at the raison d'etre of this outback town. The weary traveller, cold and hungry, probably fails to notice them, and makes his way as quickly as he can from the station to his hotel room in nearby Argent Street.

But next morning the clear light of day reveals the true nature and extent of the amazing "Silver City." The main streets are reminiscent of any large, Australian country town, but more distinctive is the extensive corrugated and ripple-iron suburbia. And utterly unique is the back-drop on the southern side, a massive continuous heap of grey slag along the "line of the lode," stretching for several miles and interspersed here and there with the head-frames, power-houses and other mighty paraphenalia of the mining industry. Going deep underground in the silver-lead-zinc mines of Broken Hill is a stirring enough experience, but equally dramatic is the view from the plat half-way up a head-frame - out across the slag and mills, over the sprawling township and beyond its especially-planted ring of trees to the vast, endless plains of red sand - the uniformity relieved only by the patches of stunted vegetation and the pimple-like hills of the surrounding ranges. In all Australia, there is no-where quite like the "Silver City."

During the 1880s, Cornish immigration into Australia dwindled to an all-time low, especially in South Australia where it ground to a halt with the ending of the assisted passage scheme in 1886. Indeed, for South Australia, the 1880s were a bad decade all round, and it was during those years that a major movement of Cornishmen away from the colony became noticable. The Victorian and (to a lesser extent) New South Wales goldfields had attracted many of South Australia's Cousin Jacks during the 1850s, but the majority of these had returned to the Central Colony, as the diggings became more difficult and less profitable. A fair number came back to restart the Burra, Kapunda and other mines, while the discovery of the northern Yorke

Peninsula copper deposits in 1859-61 caught the imagination of many of Victoria's Cornish miners. By the 1880s, however, Burra and Kapunda had been abandoned, along with many other South Australian mines, while Wallaroo was seemingly on the point of collapse, with Moonta faring only marginally better. Thirty years of ardent prospecting in the colony had failed to uncover any goldfields even remotely comparable with those of Victoria and New South Wales, and unemployed miners in South Australia had little hope of finding work in other industries for the 1880s witnessed both a slump in the local economy and a contraction of the northern wheat frontier as a result of the long drought.

Just as Adelaide's financial crisis of the 1840s had led to an upsurge in mineral prospecting, so the economic downturn of the 1880s caused many South Australians to search anew for mineral wealth - but not always in their home colony, for copper was by now unprofitable, and the lack of local gold had already been demonstated. Thus it was that many - most especially the Cornish - crossed the border into the Barrier Ranges of New South Wales, searching for gold and any other metal they might stumble across. (The Barrier is a bleak range of sharp, rocky hills which march curiously across an otherwise featureless land-scape, and the whole Barrier district is the arid expanse of north-western New South Wales - bordered on the west by South Australia, and in the north by Queensland).

Although obviously mineralized country (there were the usual, tell-tale metal-stained outcrops), the harsh conditions on the Barrier had acted hitherto as a deterrent to prospectors. Many remembered the ill-fated Barrier Rush of 1867, J.B. Jaquet writing in 1894 that,

The history of metalliferous mining on the Barrier Ranges may be said to have commenced this year (1867) when a rush of men in search of gold took place from Burra Burra in South Australia, which was attended with many sad results. The prospectors, wandering in the almost waterless and sparcely settled country, endured terrible hardships, and not a few perished miserably of thirst. Many stories are told by the old pioneers of the circumstances which led to this rush ... The sole desideratum of the prospectors was gold, and in their search for this metal they must have passed over the deposits of silver ore which twenty years later were to give the field a foremost place among the mineral districts of the world.

An equally grim account of the Rush was given in 1904 by Donald Clarke, who wrote that,

The discovery of gold in Victoria and New South Wales unsettled men's minds, so that when, in 1867, it was reported that gold had been found in the Barrier Ranges, miners from Burra Burra, in South Australia, rushed off to the new land of promise. The tale of the sufferings and privations of many of these unfortunate men will never be told. In an arid, semitropical (sic), and almost waterless country, many perished, while the survivors found no gold, and returned with such harrowing descriptions of the place that it was shunned as a land accursed...

The 1867 Rush was in several ways a precursor of the Barrier discoveries in the 1880s. And just as the latter were fascilitated considerably by the northward flow of redundant miners following the Burra closure, so the Rush of 1867 was provoked by the temporary suspension of operations at the Burra in the early part of that year. In February the *Wallaroo Times* reported an alleged discovery of gold on the Barrier and anticipated that "As the Burra mine will be knacked next month... we would not wonder if some of the miners made tracks for that locality" ("knacked," the Cornish mining term for "abandoned," was used frequently in the Australian press). A few weeks later it was asserted that the "discovery" was an elaborate hoax perpetrated by a former Burra miner attempting to sell a worthless claim in the area (he showed a Bendigo nugget to prospective buyers, saying that he had found it on the Barrier).

But although the evidence served to confirm that the "find" was indeed a fraudulent claim, this did not prevent "The utmost excitement..." from being generated in South Australia. In the Yorke Peninsula mining town of Kadina, "... no other subject was spoken of or talked over but the Barrier Ranges...," according to the local newspaper, and although "... many of the old hands did not seem to relish the news ...," a number of miners were preparing to leave the district. At the end of March a Cousin Jack by the name of Waters returned to the Burra to pick-up supplies. He had no gold, but he was able to describe the highly-mineralized character of the Barrier to the Cornishmen at the ailing mine. At the same time a contingent of miners left Moonta, "... notwithstanding adverse reports ..." in the local press, and at Wallaroo "The gold fever..." was also caught. But less than a month later the men were returning, none of them having found the illusive (if not illusory) gold, and one furnishing a *Wallaroo Times* reporter with the grisly details of his discovery of a decomposing body on the Barrier trail.

It was small wonder, them, that most prospectors avoided the Barrier. It was not until more than 10 years later, in 1876, that serious attention was again focussed on the district, following the discovery of galena (silver-lead) at a place called Thackeringa. A parcel of 36 tons of ore was raised and sent to the Burra, and from there to Britain for expert analysis and perusal. But somehow it was lost during its voyage, and thereafter the Thackeringa diggers lost interest - the working lying idle until the early 1880s. Then, with the onset

137

of "hard-times" - in South Australia in general, and in the mining districts in particular - prospectors yet again began to cross the border into New South Wales. They included some of South Australia's most hardened and experienced diggers, among them John Treloar from Helston (who had been a pioneer on the Victorian goldfields) and George Prout from Camborne.

Prout had also been in Victoria, and had purchased land there in 1866. But he had been unable to adapt to the pastoral life, and by 1874 was back in Adelaide where he found work as a carpenter. Tiring of this "domesticated" existence, he began again his wanderings - this time taking the Barrier trail, and pushing northwards through the Barrier Ranges to the Mount Browne district. There Prout, and others like him, encountered fearful conditions - a total lack of provisions, no water, and no-where to escape from the dreadful heat of the outback sun. Preferring defeat to death (not all diggers were so sensible), Prout made his way back to Adelaide, even though his search for gold had so far been fruitless. But he was not entirely beaten, and was appointed manager of the "South Australian Prospecting Association" on whose behalf he searched for minerals in the far-north of the colony. He again attempted to work his way through to the Barrier Ranges but, like many other luckless Cousin Jacks, was forced to return by the appalling drought conditions which had already beaten him at Mount Browne. Prout later prospected in other parts of the continent, in Western Australia and in Queensland, and he became a successful engineer and businessman - his work taking him on visits to as far away as Cornwall, Wales and Sicily. But never again did he go back to the Barrier.

Unlike the Rush of 1867, there was gold to be had at Mount Browne - but at a terrible price. Drought, typhoid and heat took their toll, Fred Blakely writing in 1938 that,

It was never known how many perished on that track, but in my boyhood old settlers on the Barrier still told stories of the Mount Brown (sic) gold-rush and disaster. Mount Brown is in New South Wales just within the north-western angle of the state, but the rush was from South Australia to this place. The nearest town was Hawker, in the north of South Australia, three or four hundred miles from Mount Brown, and a man has to take his hat off to those old diggers for blazing a track across such a country - all right as long as rain came, but a death-trap when it failed.

The return from Mount Browne coincided with renewed activity on the Thackeringa silver-lead fields, and great excitement was caused by the discovery, in that area, of the Umberumburka mine. A new township, appropriately named Silverton, sprang up around this new working in 1882 and 1883, and, although the Adelaide *Advertiser* warned that "There is nothing but death and desolation on the Barrier," a stream of South Australians poured across the border (even as late as 1891, two-thirds of the population of Silverton was from South Australia - a great many of them, as Brian Kennedy shows in his splendid book *Silver, Sin & Sixpenny Ale*, being Cornishmen and their offspring). One correspondent in the *Yorke's Peninsula Advertiser* described the half-starved condition of men who had "... tramped it from Adelaide...," while another wrote that "My advise to miners is 'don't be mad enough to go there'... Many old Moonta men that are here have helped me to these conclusions." Yet another newspaper report asserted that "...the miners' whisper is that all the Moonta men are likely to return, as they don't like Silverton a bit."

But, however harsh life on the Barrier may have been, the prospect of work - and maybe wealth - appealed to many an unemployed Cousin Jack on the Peninsula. In November 1885, for example, Captain Luke sent down to Moonta from Silverton for 20 miners and a blacksmith "... and had no difficulty in securing the number..." while only a month before the *Register* had noted that many were leaving Yorke Peninsula for the Barrier silver-lead fields. "Silverton News" became a regular feature in Moonta and Wallaroo newspapers, the articles detailing the activities of Cornish captains such as Captain Phillips on the Umberumberka Extended and Captain Williams in the West Umberumberka. In far-away Cornwall there was also some interest in the Silverton field, an article in the *West Briton* in 1884 explaining that,

Advices received in Camborne state that the new silver field opened up near Adelaide has proved to be very rich. Scores who went to the place poor men were in six weeks rich. The diggings prove to be a great source of employment, and no doubt to many Cornishmen.

It is interesting to note the comment "near Adelaide," reflecting as it does the *West Briton's* literal acceptance of its Australian correspondent's typically casual sense of distance - as far as the latter was concerned, Silverton was virtually in South Australia and therefore by definition "near Adelaide!" Be that as it may, the *West Briton* was certainly correct in its assumptions. In the depressed atmosphere of the 1880s the silver fields were valuable sources of employment, and of course they attracted many Cousin Jacks - the local Baptismal Register, for example, recording the names of many Cornish parents, such as William and Emma Bray, Thomas and Jane Poole, Henry and Ellen Hancock, and William and Alice Roberts.

Like many embryonic mining camps, early Silverton was a "rough-and-ready" place. But the "civilising" effect of the South Australian Cornish was an important one, the Revs. John Thorne and Charles Tresise crossing the border to establish a preaching place. Thorne, in fact, had been born just inside Devon and was not therefore a Cornishman. But, like Captain

Hancock, he was "near enough to one" - he was a member of the celebrated Thorne family which had helped William Bryant (or O'Bryan) establish the Bible Christian denomination in the Cornwall-Devon border districts in the early nineteenth-century, and he had also worked amongst the Cornish miners on Yorke Peninsula. Charles Tresise was a Cousin Jack through and through. He was born at St. Erth in 1843 and, according to one Methodist obituary, arrived in Australia... "in early manhood... where he gained much public recognition for his manliness, intelligence, and high Christian character...," entering the Bible Christian ministry in 1863. He worked first of all at Kapunda, but was most especially known for his pioneering efforts on the Barrier. Together Thorne and Tresise, along with their lay followers, brought their Cornish brand of Methodism to Silverton, and it was not long before, as Geoffrey Blainey has said - one could hear "...the voices of Cornish men and women singing hymns and part-songs as they filed over the hill from the Bible Christian chapel to the Umberumberka mine."

Despite all the activity and the extent of the prospecting, Silverton only ever produced two really worth-while mines - the Day Dream and the Pinnacles - and the hey-day of the town was relatively short, roughly from 1884 to 1886. And even more important than the rise of Silverton in precipitating the exodus of Cornishmen from South Australia to the Barrier was the discovery of Broken Hill in September 1883 by Charles Rasp ("German Charlie"), a boundary rider from the nearby Mount Gipps station. The "Hill" itself, from which the "Silver City" took its name, was in fact a typical rocky, mineralized outcrop - situated some 20 miles east of Silverton, but an integral part of the Barrier silver-lead field. Rasp thought he had found tin, but it was of course galena, the first ore assaying at some 700 to 800 ounces of silver to the ton.

Along with Rasp, perhaps significantly, on the day of the discovery was James Poole, a Cornishman from Kapunda who was then also working at Mount Gipps. Poole was born in Cornwall in 1848 and had worked in South Australia before going to the Barrier. Together with Rasp and their other colleagues from Mount Gipps, Poole helped form the famous "Syndicate of Seven," the original owners of the Broken Hill shares. Poole, however, grew tired of waiting for the mine to prove a bonanza, or perhaps he believed it never would be, and he disposed of his interests in the syndicate during 1884 and 1885 - before they had realised their true value. Thereafter, he lived variously at Broken Hill, Burra, Kapunda and in Western Australia, before retiring finally to Kapunda, where he died in 1924. Another of the original "Syndicate of Seven" also threw away a potential fortune, losing his shares by staking them in the celebrated playing-card "Game of Euchre," an event which has passed into Australian mining mythology. It is rather fitting that men should have gambled for mining shares by playing euchre, for euchre was so common in Cornwall and so rare elsewhere that it might almost be considered "a Cornish game!"

Little work was done at Broken Hill until towards the end of 1884. But then an advertisement in the *Silver Age* newspaper drew a number of copper miners, mostly Cornishmen, to the Hill. Samuel Sleep, orginally from Moonta, was appointed Captain, and in the October a Cousin Jack by the name of Rosewarne began sinking Rasp's Shaft - at the tutwork bargain of 45 shillings per foot. New samples assayed as much as 2,000 ounces to the ton, and during 1885 the syndicate reformed itself as a company - Broken Hill Proprietary (BHP). Thereafter, development was rapid, both financially and in terms of physical expansion, and by 1887 BHP shares with a nominal value of £19 were fetching as much as £174 10s. In February 1888 they went as high as £409 each.

Soon other mines were started - Broken Hill North, Broken Hill South, the Victoria Cross, the Great Northern Junction - and before long the whole "line of the lode" was bristling with silver-lead workings. The rapid expansion of the Broken Hill mines increased the rate of immigration from South Australia, and the bond between the Central Colony and the Barrier became even more intimate. In 1888 one observer wrote that Broken Hill "... geographically belongs to New South Wales, but commercially to South Australia..." Adelaide was, he said,

... by long distance the nearest city, and Port Pirie its nearest port. The great majority of the people are South Australians, with families and friends in South Australia, and each desire to trade with and benefit (other things being equal) South Australia; and ... much of the capital expended in the district is South Australian...

He might have added that a large proportion of these South Australians were Cornishmen. Many were miners from Moonta and Wallaroo, others were men from Kapunda and Burra who had been indifferently employed since the closure of their mines, and still others were Cousin Jacks from the various centres of population such as Adelaide and Gawler. In the early 1880s, Terowie was the most northerly railhead in South Australia, making the journey to the Barrier a long and difficult one. But in 1885 work began on an extension line to Cockburn on the border, and in October 1887 the author of an article "Six Months at Broken Hill by a Moonta Boy" in one Peninsula newspaper was able to describe how he had travelled straight through from Moonta to Cockburn by train. In time, the "Silverton Tramway" linking Cockburn to Silverton and later Broken Hill was opened, and the improving communication system facilitated still further migration from South Australia. There were, of course, those who "... found the Peninsula, bad as at is, better than the silver region...," while some at Moonta were deterred by the experience of one John Hamblin who contracted rheumatic fever on his way to the Barrier in 1886 (all the "gory details" were reported with great relish and gusto in the local press). Others may have heeded "Old Moonta Miner's" dire warnings in the press about the threat of lead poisoning, especially after one Cousin Jack at the Umberumberka had declared that "I would rather work in the old Moonta for 30s a week than here for £25."

But despite all this, the exodus continued. At first it was the unemployed who went, but by 1895 H. Lipson Hancock was noting the shortage of good hard-rock miners on the Peninsula as a result of "... men ... leaving the district for Broken Hill..." To some extent the Moonta and Wallaroo company adopted a policy of "if you can't beat 'em join 'em" by sending groups of its own prospectors to the Barrier district and by purchasing local claims. Most of those forsaking the Peninsula for the "Silver City," however, did so on their own initiative. An early departure was John Carthew, from Moonta, who went to Thackeringa in the early 1880s, working first in Silverton and then at Broken Hill, and becoming finally Inspector of Mines for the colonial government of New South Wales. Another was Charles Pyatt (known affectionately as "Camborne Charlie"), who had arrived in Wallaroo Mines from Cornwall in 1883, as part of the group recruited by Captain Richard Piper, but was soon enticed to the richer fields of Broken Hill. Others made their way up to the Barrier from Burra Burra, the *Burra Record* in February 1890 noting that there were many "... old Burra boys..." in Broken Hill. Indeed, as late as 1905 it was said that there were as many as 300 Burra men on the Barrier, one report asserting that:

We are gradually losing our (Burra) population, the place of refuge being Broken Hill... The other night no less than thirty Burra boys were seen in Argent Street, quietly conversing together, and some distance further on another "bunch" was seen.

The influxes continued, and by 1891 there were some 20,000 people huddled together in the ripple-iron town of Broken Hill, the "Silver City." Early buildings, many little more than shacks, had been erected hurriedly Moonta-style by the miners, but the town was by now aquiring gradually an air of permanence. Argent Street, Broken Hill's commercial centre and main thoroughfare, had gained a few stone-built pubs and shops - with verandahs to boot - and a more disciplined approach to building (with roads organised on the typically Australian grid pattern) was by now in evidence. The street-names themselves read like terms straight from a metallurgy text-book - Argent, Sulphide, Oxide, Blende, Carbon, Crystal, even Slag Street!

Conditions remained poor, however. Broken Hill continued to suffer the effects of its isolation - goods and services were relatively hard to come by, and tended to be expensive. And surrounded totally by arid, dangerous, semi-desert, the town was almost a prison for many of its inhabitants, who sought escape from the harshness in the pubs or in the chapels. Water supplies were always a problem, the district often subject to drought and fearful dust-storms, and typhoid (known locally as "Barrier Fever") took its toll, as it did in most mining camps. In 1904 Donald Clark had this to say of Broken Hill:

In the summer the temperature rises to over 100 degrees in the shade for days together... not one shrub is to be seen for miles round the Hill, and the scattered tufts of saltbush left alone redeems the place from being a desert... Looking down on the town from the hill, one sees fine buildings and streets, but the background of low hills and bare ridges makes the place look desolate... The bare houses, with their baked backyards, and the hopeless attempts made to grow a few common flowers, make the place a very unenviable one to live in, in spite of the brave fronts of many of the buildings and shops.

But such harshness and austerity was not likely to stop the more dedicated brand of Cousin Jack. And a considerable number of those attracted to Broken Hill in the 1880s and 1890s were typical of the itinerant prospector-cum-digger class, and had already spent many years roaming across Australia in search of mineral wealth. There was Joseph Kemp, from Redruth, who "... worked on about every mining field in the Commonwealth, and went to several gold rushes ...," and Stephen Pellew Carthew - also born in Redruth - who had toiled in more than half-a-dozen mines in South Australia and Victoria before going to Broken Hill. William Matthews, later to become the South Australian Inspector of Mines, travelled down to the Barrier from Darwin in the Northern Territory (where he had been prospecting for tin), while St. Austell-born John Rowett crossed the border into New South Wales after some time working in the Mutooroo and other "far northern" mines in the Central Colony.

The Wesleyan Marriage Register for Silverton and Broken Hill in the 1880s and 1890s reveals the extent of the Cornish-South Australian element in the local population. The marriages themselves were solemnized by the Revs. Henry Trewin, Thomas Trestrail, and John Grenfell Jenkin (all three of whom had been active in the Central Colony), while those joined in holy matrimony included: Robert Hooper from Kapunda and Elizabeth Triplett from Moonta, John Rowe from Adelaide and Catherine Roberts from Moonta Mines, David Isaac from Kapunda and Emma Hall from Devon, William Kellaway from Kooringa (Burra) and Mary Jane Roberts from Moonta, George Snell from St. Just and Janet Pate, William Oliver from Clare and Susan Tozer from Stithians, William Scown from Moonta and Catherine Pergam from Scorrier, John Hocking from St. Blazey and Elizabeth Whitburn from St. Cleer, Alfred Chappel from Moonta and Catherine Harvey from Camborne, John Mitchell from Kadina and Beatrice

Webster from Kadina ...

Local business directories, while not giving place of birth or earlier residence, also reveal more than a smattering of Cousin Jacks. At Thackeringa in 1891 there were miners with names such as Cornish, Davy, Edwards, Polkinghorne, Sampson, and Williams, while at the Purnamoonta there was a blacksmith called Trewick. The Pinnacles, a settlement named after the mine of that name some 10 miles south-west of Broken Hill, could boast a Britton, a Berryman, a Brokenshire, a Bennetts, a Dunstan, several Hoares, a Martin, a Mitchell, a couple of Pearces, a Peters, a Rosewarne, a Rowe, a Richards, a Sibley, a Trembath, a number of Tremaines, and a whole colony of Thomases. The Trembath was probably James Warren Trembath. He was born in St. Just-in-Penwith in 1836 and migrated to Victoria during the Gold Rush days to mine at Bendigo. He soon moved on to South Australia, however, where he found work in the Wallaroo Mines, and lived at Kadina until 1891 when he went to the Pinnacles. He died there in the mine in 1902, from heart failure.

A closer examination of the numerous mines opened-up in the Barrier Ranges before the turn of the century again lends evidence to illustrate the important Cornish-South Australian influence. In June 1886, for instance, the *Yorke's Peninsula Advertiser* noted that the manager of the large Junction mine was Captain George Rogers, a man known for his "... experiences in Cornwall and his subsequent experience here (Moonta)." A few months later the same newspaper reported that the Round Hill mine was being opened-up by Captain Matthews, and that the celebrated Captain Richard Piper - formerly of Wallaroo Mines, the man who had undertaken the famous recruiting trip to Cornwall in 1882 - had been appointed underground captain at the Broken Hill Proprietary mine, where Samuel Sleep was already manager. Piper was later associated with both the North and South mines, while James Retallick was at the British, and James Hebbard - well known at Bendigo and at Wallaroo - was at the Central. John Warren, who had earlier run the Hamley Mine at Moonta, became one of the most popular of Broken Hill's personalities during his term as manager of the Block 10 mine, his underground captain there being Dick Thomas - another "Moontaite." The Victoria Cross mine was managed by yet another Cornishman, William Kerby, who was born in Liskeard in 1854.

Perhaps the most important of the earlier Barrier captains was William Henry Morish. Morish was born in Truro in 1844, at the age of 16 migrating to South Wales where he spent two years working in the coal and iron mines. In the early 1860s he went to South Australia, where he found employment in the Wallaroo Mines. From there he travelled on to Bendigo and Ballarat in Victoria, and to Great Cobar (where he was underground captain) in New South Wales, moving to Broken Hill in 1886 as captain of the South and Central mines. He was instrumental in the foundation of the local branch of the Mining Managers' Association of Australia (the bosses' answer to the workers' unions), and became something of a local capitalist - dabbling in the shares of the various Barrier mines. A shrewd investor, it was not long before he had amassed enough money to retire to his villa in the Adelaide suburb of Plympton.

According to popular tradition (of which there was plenty on the Barrier), the important Junction mine was first discovered by a Cousin Jack called Penglase, while a considerable amount of early prospecting work at Broken Hill was undertaken by the "Devon and Cornwall Syndicate" - a body whose activities were a source of great interest at Wallaroo and Moonta. Many of the claims in the districts surrounding Broken Hill were also worked and managed by Cornishmen, one observer in 1888 writing that mining captains "... renew on the Ranges friendships made in Cornwall and California, New Zealand, Tasmania, and other parts of Australasia." Captain T. Tregoweth, well-known in South Australia, ran the Rise and Shine mine 15 miles north-east of Broken Hill, while Captain T. Rowe "... formerly of Yorke's Peninsula..." managed the Hidden Secret, near Silverton. Captain Tresize, who had earlier run the Talisker silver-lead mine in South Australia, was placed in charge of the Bonanza property, a working that never really lived up to its optimistic name. From Fullerton, near Adelaide, came Captain Penberthy - a man with "... a good reputation and a mining experience of thirty years..." - who managed the Big Hill Mine, while the nearby Anaconda was run by Captain Dunstan, another of Great Cobar's Cousin Jacks. Dunstan was also at the Rising Sun, while Captain Hawke was at the Lady Bevys and the Hidden Treasure, with Captain Stevens at the neighbouring New Mile claim and Captain Bennett at the Rockwell Paddock. In the Thackeringa district, Captain Polkingthorne (sic) was manager of the Gipsy Girl, while the Terrible Dick mine was named after a certain Rickard Tonkin - a local celebrity who ran the mine until 1886 when, selling the working to a Sydney firm who proceeded to spend £26,000 on its rapid development, he was replaced by Captain Hocking. Nearby was the Eagle Hawk mine, managed by Captain James Eddy, the Great Britain (run by Captain Hicks), and Captain Ellis' War Dance claim.

Some 50 miles north of Broken Hill were the Euriowie tin-fields, first opened-up circa 1888 in the aftermath of the Broken Hill discoveries, when diggers began to look further afield for mineral wealth. In 1888 Captain William Oates, from St. Just, a man with "... considerable experience in tin-mining in Cornwall...," commenced operations at the Victory Mine, while in 1889 Captain William White (a former manager of St. Just United in Cornwall and the West Bischoff in Tasmania) arrived from Adelaide to undertake prospecting work on behalf of a South Australian syndicate. Captain W. Williams, who had also worked tin in Cornwall, managed the Trident mine, and Captain Thomas - an "... eminent tin miner..." from the Dolcoath mine in Camborne - arrived in the district in 1888 at the request of the New South Wales government to report on the scope and prospect of the

145

Euriowie fields. Thomas argued against the tin-fields becoming what he called a second Gwennap (i.e. a proliferation of small claims, none large enough to be viable in the long-term) because this would lead to their premature abandonment, but was nevertheless impressed by what he saw and could not help comparing Euriowie with Dolcoath, Great Wheal Vor, and mines in the Wendron district. No-one took any notice of Captain Thomas' advice, but as if to echo his confidence, a number of the mines were given Cornish names - Wheal Byjerkno bore a Cornish prefix, while the Carn Brea was "... named after the celebrated Cornish mine...," and Botallick (sic), Dalcooth (sic), Tincroft and Mount Tincroft were drawn from examples in Cornwall.

As in South Australia, the Cornish influence on the Barrier went deeper than mining terminology and personnel. In the early days at least, the Cornish technological contribution was at least as significant as in Victoria, or even the Central Colony. Especially important was the heavy engineering plant manufactured by May Brothers' Foundry in Gawler, a large town to the north of Adelaide. The foundry was set-up in July 1885 by Frederick and Alfred May, the former being the senior partner. He was born in Perranzabuloe, and at the early age of 23 was appointed Chief Engineer at the Moonta Mines, a position he held for 11 years. In 1873 he and his younger brother Alfred went to work in Martin's Foundry at Gawler, and in 1885 they branched out on their own. According to one contemporary local history, they progressed rapidly "... from strippers and seed growers to jigs and boilers and from jigs and boilers to smelting and crushing and pumping and winding plants...," the impetus to expand being of course the rise of Broken Hill. They built machinery for the "Cornwall Smelters" at the South Mine, they constructed the concentrator and the crusher at the Umberumberka, and they manufactured the whim engine for the Britannia mine. They also produced much of the plant for the Broken Hill Proprietary company's smelting works at Port Pirie. In December 1897 Frederick May died, but Alfred kept the business going, so that by 1901 the foundry was producing the largest mining engines ever built in Australia.

As far as Broken Hill was concerned, perhaps May Brothers' greatest achievement was the successful development of machinery to deal with the so-called "sulphide problem." By the 1890s the Barrier boom was beginning to fade, with silver losing its position as an international currency. Moreover, the mines were becoming deep, and the metals were at depth appearing as sulphides - which were difficult to extract. Eventually, a chemical flotation method (still employed today) was devised, but initially the sulphide separation was by mechanical means - using a seperator invented by May Brothers. They demonstrated to Captain Richard Piper in an experiment at the South Mine that separation was possible, and at the Block 10 mine Captain John Warren, a man with considerable inovatory skill who had come to Australia armed with first-class testimonials from Captain Charles

Thomas of Dolcoath and Captain Josiah Hitchins of Devon Great Consols, was successful in applying May-built machinery to cope with his sulphide problem.

All manner of other machinery and equipment was aquired from South Australian foundries (a number run by Cornishmen), and other items were obtained from Cornwall direct - in 1908, for example, it was noted that pneumatic rock-drills manufactured by Holman's of Camborne were in common use at Broken Hill. The Cousin Jacks on the Barrier had also a thing or two to teach miners from other parts. Compared to the relatively stable ground of Wallaroo and Moonta, the Broken Hill workings had a marked tendency to "creep" and "run together." This made deep mining dangerous, but "... an old Cousin Jack precaution in unstable ground was to watch the migrations of underground rats," says Geoffrey Blainey. When the rats left a stope or a level, so did the Cornish! At first, extractive work in the Barrier mines was according to the usual Cornish open stope type, using either overhand or underhand stoping. In the late 1880s, however, the introduction of "square-set" timbering in the mines represented a radical departure from Cornish practice, a change reflecting the arrival of powerful American mining engineers and managers on the Broken Hill scene.

By that time Cornwall had lost its pre-eminent position in the mining world, and so it was only natural that the bosses of the massive and still-booming American mines should overshadow the by now eclipsed Cornish captains. But specifically, in the case of Broken Hill, the introduction of Americans to work the larger mines derived from the fact that most of the Cornish captains on the Barrier had aquired their experience mining copper in South Australia and were unfamiliar with the techniques and problems of winning silver-lead-zinc ore. Ironically, though, the first "Americans" were Comstock miners from Nevada and may well have included Cornishmen in their ranks (Captain Retallick, one of the newcomers, was certainly Cornish), as the Comstock silver workings were well patronised by the Cornish in America.

Whatever the case, the arrival of the Americans could not detract from the importance of the Cornish pioneering work in the district. The "Yanks" did not have all the answers (they exaggerated "creeping" by erecting heavy plant over unstable ground), and the Cornish had already, of course, made an enduring impact upon local society. Much of the Barrier's commercial life, for example, was managed by Cousin Jacks. Samuel Gray, born in Cornwall in 1849, arrived in Silverton from Adelaide in 1886 to establish a building firm (he constructed engine-houses for the Central Mine and the Zinc Corporation's workings), while James Burnard - born at Boscastle in February 1837 - came at about the same time from Terowie, a rural township in the mid-north of South Australia. Sydney Trenomon arrived in Broken Hill from Moonta to open a general-store, while Ernest William Crewes travelled up from the Burra to open a branch, on the Euriowie tin-fields, of the trading firm of which he was part-proprietor. Crewes came from Truro, in Cornwall, and had been initially an employee in Samuel Drew's "Cornwall House," the general-store at Kooringa (Burra) but had soon been offered a partnership. Thereafter, the business blossomed (branching-out to the Barrier was a good move), Ernest Crewes becoming Mayor of Burra Burra in 1900-02. George Sara, born in Perranwell in 1839, followed Crewes' example and established a branch of his Burra building company (Sara & Dunstan) at Broken Hill, while St. Just-born Richard Warren moved from Wallaroo Mines to set up a store in Oxide Street. Other local traders - such as John Mitchell (ex-Kadina), Pellew & Moore, Penhall Roberts & Co., J.W. Pengelly, and R.J. Hooper & Co. - also exhibited an obvious Cornish origin.

Perhaps more importantly, the path of Barrier social history - at least in the first two decades - was to a considerable extent moulded by the Cornish-South Australian experience. The Cornish cultural impact was manifested in various ways, despite the fact - as Blainey has shown - that Broken Hill was never a Cornish mining town in the way that Moonta and Burra were. There was, however, a suburb of south Broken Hill known as "Moonta Town," and local hotel name such as the "South Australian" and the "Duke of Cornwall" were more than a clue to the district's cultural antecedents. The Cornish were well-known for their mining proverbs and expressions, and in 1888 one visitor to the Barrier recorded a number of miners' sayings that were in use there - "A good prospector is always sanguine," "Great gains are never without the companionship of losses," and so on.

Although only marginally different, in a physical sense, from other Britons and Australians (they tended to be dark and stocky, and spoke with a distinctive accent), the Cornish were somehow easily recognised by other settlers. The following story, recorded by Blainey, shows the extent to which this was so: At Broken Hill Proprietary Mine a certain William Harry supervised an overhead shute which loaded ore into trucks in a level, underground. Some thought this dangerous, and Harry was questioned about its use by government Commissioners in 1897. The ensuing conversation was most revealing:

Q. Is it called all sorts of names?
A. Yes, "Cousin Jack," "Irishman," and such like, just according to what the navvy takes it into his head to call it. He generally names it after the man on whom he has the greatest set; and I suppose that is the reason why it has been named here "Chinaman."
Q. You want competent men, and you do not care what nationality or association they belong to?
A. No; I would not care if they were Chinamen; but I bar "Cousin Jacks" sometimes.
Q. Very good; but would you ask them if they were "Cousin Jacks?"
A. No; I know them by sight.

One of the first cottages built at Moonta Mines, erected by a Cornish miner named Peter Bowden

**Sir James Penn Boucaut, from Mylor, the miners' friend, South Australian Premier, High
Court Judge, and Acting Governor, Chairman and President of the Cornish Association of
South Australia at its foundation.** Courtesy of S.A. State Library

Strike leaders and their supporters in the "Great Strike" of 1874 at Moonta and Wallaroo
Courtesy S.A. State Library

"Forest Creek", Victoria - once the home of many Cornishmen. Sketch by W.A. Nicholls from John Sherer, The Goldfinder of Australia, 1853.

Courtesy Penguin Books (Australia)

Nicholls' "Diggers Executing Their Own Laws" - the rough justice of the Australian goldfields.

Courtesy Penguin (Australia).

Nicholls' "Successful at Last"   Courtesy Penguin (Australia)

154

"The Unlucky Digger That Never Returned" - Nicholls illustrates the fate of not a few **Australian prospectors**

"The Invalid Digger". "Misfortune, disappointment, and sickness had done their work upon him" say Nicholls and Sherer.

"Preaching At The Diggings" - Nicholls' impression of the Cornish lay-preacher, John Trevellyan, at Forest Creek
Courtesy Penguin (Australia)

157

The "Welcome Stranger" nugget, pictured with its Cornish discoverers - John Deason [with crowbar] and Richard Oates [with pick] - at Bulldog Gully on the Victoria Goldfields, 1869.

Courtesy La Trobe Library

Argent Street - the principal thoroughfare of Broken Hill, the "Silver City" - looking North from Billygoat Hill in 1886

Courtesy Charles Rasp Memorial Library

159

A view along "the line of the lode", looking south from the township, showing the giant stack of the Broken Hill Proprietary Mine, late 1800s.

Courtesy Charles Rasp Memorial Library

160

The Great Boulder Mine, Western Australia, in 1896. Kalgoorlie Western Argus

The Great Boulder Mine, Western Australia, in 1896. Kalgoorlie Western Argus

Under Her Majesty's Commissioners.

ENTIRELY FREE
EMIGRATION
TO
VAN DIEMEN'S LAND
AND
New South Wales.

Mr. LATIMER,
OF TRURO,

Is desirous of obtaining, IMMEDIATELY, a LARGE NUMBER of Emigrants belonging to the class of *Mechanics*, Handicraftsmen, *Agricultural Laborers*, Carpenters, Quarrymen, Masons, and Domestic Ser ants.

The Emigrants must consist principally of married couples. Single wom , with their relatives, are eligible, and in certain cases, *single men.*

The age of persons accepted as Adults is to be not less than 14, nor, generally speaking, more than 35 ; but the latter rule will be relaxed in favour of the parents of children of a working age.

The Colony of Van Diemen's Land has been established more than half-a-century, and possesses the usual advantages belonging to the Australian Settlements. It is not subject to drought, and affords a peculiar demand for the classes above-named.

No CHARGE for CHILDREN!!!

Applications, *post-paid*, or personal, to be made to Mr. LATIMER, 5 Parade, Truro.

E. HEARD, PRINTER AND BOOKBINDER, BOSCAWEN-STREET, TRURO.

"Entirely Free Emigration to Van Diemen's Land and New South Wales" - yet another of Isaac Latimer's posters   Courtesy Royal Institution of Cornwall

TO
MINERS,

Masons, Carpenters, and all persons engaged in building,
Blacksmiths,

AGRICULTURAL
LABORERS

And Mechanics generally.

Eligible married persons under 35 years of age, may have

AN EARLY FREE PASSAGE
TO
NEW
ZEALAND

ON APPLICATION

To Mr. LATIMER, Truro.

Married people above 40 years of age, may have a *free passage* if they have one child 14 years of age and upwards, for every year that the parents exceed 40,---that is, a father 42 years of age, or about, will be acceptable if he has 2 children above 14.

"To Miners... An Early Free Passage to New Zealand". Isaac Latimer acted as migration agent for the most far-flung corners of the Antipodes

Courtesy Royal Institution of Cornwall

"Honest John" Verran, from Gwennap, South Australia's first Labor Premier

Courtesy S.A. State Library

Many of these Cousin Jacks were real "characters," some to the point of eccentricity. There was, for example, Richard Thomas, the underground captain at the Block 10 mine. One story tells how he came across a group of miners loafing. "Are you new men,?" he asked, "I think the sun has been shining on your heads all your life," meaning that they were not proper underground workers. Later he heard the group whistling and singing, and so dismissed them from the mine's employ, explaining that "No man can strike a drill into hard rock, and whistle and sing at the same time. It takes a man all his wind to fulfull his duty."

Another great Barrier "character" was Captain Richard Piper, and yet another tale - this time related in 1889, by W.R. Thomas - lends an insight into both his own "Cornishness" and the survival of traditional Cornish lore at Broken Hill. Apparently, one day a miner approached Captain Piper, requesting a job. Piper replied by asking the applicant if his name was Oates. Somewhat startled, the man answered that it was, and Piper explained his foreknowledge by revealing that he had noticed the miner's stunted fingers. Piper then went on to recount, so the story goes, how 150 years earlier the Oates family had been notorious "wreckers" in Cornwall, plundering vessels that had had the misfortune to be driven ashore on the Cornish coast. On one occasion a ship was wrecked, and the sole survivor clambered up the cliff-face. As he reached the top, however, a wrecker named Oates cut off his fingers, causing him to crash to his death on the rocks below. There would now be no survivors to witness the plundering, but then Oates found to his horror that the sailor he had killed was none other than his eldest and favourite son. Filled with remorse, he plunged himself into the raging sea and was drowned. The other Oates sons then left the district, thereafter living a life of Godliness in Truro. But ever since, according to Piper, male members of the Oates family have been born with stunted fingers on their right hand.

Apocryphal or not, that the story should have survived at Broken Hill is in itself significant. Other cultural survivals included the formation of brass bands - the first was formed by a Cousin Jack called Kendall - and the practice of Cornish Wrestling. One of the greatest matches was the Barrier Championships of December 1890, long remembered with nostalgia in the "Silver City," and a celebrated local champion was Jacob Burrows - a Cornishman who was born in St. Austell in 1838 and had arrived in Broken Hill in 1887 from the Burra. The pervasive quality of the Cornish inheritance was also demonstrated in 1901 when the *Barrier Truth* newspaper carried a short-story "Union by Love" which, as well as playing upon a Christian Socialist theme, involved fictional characters such as Marion Tregellas, "... a Cornish Barrier lassie...," and a "... jovial young Cornishman, Charles Pentreath." It was also evident in the creation of the Cornish Association of Broken Hill in 1892, the *Burra Record* noting proudly that, "In a large mining field like Broken Hill, as with all other great mineral centres, it is not surprising to find the Cornish element is so predominant, for Cornishmen and mining are closely allied.". The Association existed for "... mutual pleasure and profit...," while its committee very liberally "In order to make the gatherings more attractive... decided to admit lady members..." In 1892 John Dunstan, formerly of Burra Burra, was appointed President of the Association.

In less self-conscious ways, the Cornish influence was also exhibited. There were any number of Cornish-Australian "Cousin Jack yarns," a number revealing the fatalistic humour of the Cornish miner. Geoffrey Blainey records that,

The story, in several versions, is told of the man who offered to break the news to the wife of a miner who had just been killed. A dour and forthright man, he went to the dead miner's house, and his heavy knocking summoned the wife. With a melancholy air he took off his hat and said: "Good afternoon Widow Tregonning." She immediately bridled: "I'm no widow. My husband's down there working." "Would you like," said the news-breaker, "to take a bet on it"

Fatalism was an attitude of mind encouraged by the Methodist outlook - God looked after His own, but if a miner was "taken" it was an expression of His Will. A popular hymn at Broken Hill was the one that went:

> Far down in the earth's dark bosom
>> The miner mines the ore:
> Death lurks in the dark behind him
>> And hides in the rocks before.
> Yet never alone is the Christian
>> Who lives by faith and prayer;
> For God is a friend unfailing,
>> And God is everywhere.

As in Cornwall and South Australia, a great many of the miners were Methodists. The first-ever religious service at Broken Hill was preached by Gwennap-born George Henry Paynter. A son of Captain William Paynter, George had arrived in Australia as a young lad and grew up at Moonta. He entered the Bible Christian ministry and, as well as being one of "... the first to hold services and build churches in towns and neighbourhoods ..." in the far-north of South Australia, he toiled on the Barrier where - according to one contemporary report - he "... did pioneering work in every part of that extensive Silver region in the early days." Others included the Rev. James Trewin, a Bible Christian who had worked in Kilkhampton circuit before migrating to Australia, and the Rev. John Grenfell Jenkin. Jenkin was born at Penzance in August 1865, emigrating with his parents to New South Wales

166

in 1877. A devoted Methodist, he was ordained and sent first to Renmark, in South Australia, and from there to Broken Hill where he gained invaluable experience of work amongst the miners - experience which was to prove its worth when he was transfered to the Kalgoorlie goldfields in 1897. Jenkin enjoyed an enormously long career, dying in Adelaide at the age of 101 in 1967.

Of course, the Methodists relied very much upon their laity - the tireless local preachers and class leaders. James Bennetts from Camborne, who had held the first religious service at Moonta Mines, helped found the Blende Street North Bible Christian chapel in the mid-1880s, and at the Day Dream mine Methodist anniversary in 1886 it was noted that the officials and choir were "... composed chiefly of Moontaites...," the tea-treat organisers being "... mostly from Moonta." In 1887, Jacob Burrows (the wrestling champion) arrived in Broken Hill to establish the Blende Street Primitive Methodist chapel. He, too, was not an ordained minister but instead a "hired local preacher," and was a great success with the people he had been sent to tend - it being said in 1888 that Burrows was "... an effective preacher to the Cornish miners." His iron chapel had seating for 150 people, and his congregation included many Cousin Jacks and Jennies, such as Eliza Uren from Calstock and Wallaroo Mines, George Trenerry from Newlyn East and Moonta, and John Bishop from Redruth and Moonta.

Although one commentator wrote that "... it is the exception and not the rule for miners to go to Church at Broken Hill...," it is reasonably clear that the chapel-going section of the population was drawn very much from the Cornish community. At the Roman Catholic Church in 1888 "... the attendance is not very large." But, in contrast, the Bible Christian chapel was big enough to seat 400 people and had an average attendance of some 170 to 200 persons, together with a vigorous Sunday School. The Wesleyans, whose minister (the Rev. A.J. Fry) had been transferred from Port Adelaide, had room for 200 people, but were busy building a new chapel, and both the Baptists and Congregationalists were strong. It was also significant that "... interesting theological discussions may be heard, started by miners, who, objecting to dogmas, are nevertheless dogmatic" - how very Cornish that sounds!

Moreover, the roll book for the years 1892 to 1912 at the South Broken Hill Wesleyan chapel indicates that the Cornish were clearly in the majority there, as they surely were at the other Methodist meeting places. Although in 1892 Brother and Sister Prisk were recorded as "Gone to Moonta," many others remained to become stalwarts of the chapel, their surnames revealing their Cornish origins: Dunstone, Scown, Hawke, Angove, Pascoe, Treloar, Bray, Rundle, Tremelling, Pryor, Rowse, Goldsworthy, Trebilcock, Angwin, Johns, Davey, James, Magor, Nancarrow, Rogers, Tredinnick, Sandow, Prideaux, Osborne, Berriman, Nankivell, Kellaway, Odgers, Hicks, Moyses, Sara, Tresize, Pollard, Gilbert, Bastian, Penna, Lander, Trenberth, and many, many more. There were the usual "tea-fights," anniversary marches, Revivals and other activities associated with Cornish Methodism, while in 1889 W.R. Thomas could write that,

... the finest sight that catches one's eye of an evening in Broken Hill is the gas-lit Wesleyan Church that stands out with all the solemnity of an English cathedral - a lasting monument to the home training of our Cornish friends.

The Methodist influence, as in Cornwall and South Australia, was also felt in a variety of other institutions. The Temperence movement, for example, was strong. John Isaacs, from St. Austell, had lived at Kapunda, Moonta, and Kadina before going to Broken Hill, where he became involved in the various Temperence organisations: "No beer in the blood" was his favourite expression! However, a perhaps more important influence was that of the nonconformist-guided Cornish Radical Tradition in the development of trade unionism on the Barrier. Brian Kennedy has noted the impact of the moderate, liberal trade unionism the Cornish miners brought with them from South Australia, and he has also pointed to the underlying influence of Methodism in the movement on the Barrier. One of the first union agitators, for example, was a certain William Rowe - a Wesleyan miner from Moonta - and Josiah Thomas, who became President of the Broken Hill branch of the Amalgamated Miners' Association in 1892, was also a Cornish Wesleyan local preacher.

The Barrier miners began to organise as early as January 1886 when, according to the *Yorke's Peninsula Advertiser*, a branch of the Amalgamated Miners' Association (AMA) was formed at the Purnamoota mine. By May of 1887 there were already signs of friction between miners and management, the men being worried about the effects of plumbism - or "getting leaded" - and expressing hostility to attempts to introduce a tribute-style contract system. The *Yorke's Peninsula Advertiser* wrote that,

A number of old Moonta miners and others are having some difficulty with the management of the Broken Hill Mining Company. The difficulty there as here appears to be about the contract system, which is not popular with the men up there any more than in Moonta Mines.

This link between trade unionism on the Barrier and on northern Yorke Peninsula was made all the more obvious by the widely-discussed suggestion in 1889 that the two districts combine to form one giant AMA branch, while in the November Broken Hill unionists wrote in support of industrial action taken by their colleagues at Wallaroo and Moonta. One Peninsula activist, Jimmy Peters, explained the link by pointing to the number of Cornish miners

168

from South Australia on the Barrier and noting that AMA "... members were continually coming and going between both towns" (i.e. Moonta and Broken Hill).

This solidarity was further confirmed by the events (as we shall see) of the "Great Strike" at Broken Hill in 1892, the first major clash between men and management, and indeed the first really violent clash between capital and labour in the Australian mining industry. The struggle of 1892 was difficult and protracted (it lasted from the July to the November), for the mine owners and captains were also organised. Unlike Yorke Peninsula where after 1889 there was only one major company, on the Barrier there were a considerable number of different, independent ventures. To ensure unity of strength and action, therefore, the captains had in October 1886 formed a local branch of the Amalgamated Mining Managers' Association of Australiasia. Captain Morish was its first President, while Richard Piper was one of its most outspoken anti-unionists.

Trouble between the opposing camps had been brewing for some time. Ralph Stokes, editor of the *Rand Daily Mail,* felt that in such an inhospitable environment as the Barrier restlessness - and thus conflict - was inevitable, but there were more tangible causes of unrest. The *Barrier Miner* (a pro-worker newspaper edited by a South Australian Cornishman called Samuel Prior) complained in March 1892 that unionists were being victimised, and unemployment and especially plumbism were sources of growing discontent. The frightening effects of lead in the atmosphere on the human constitution were noted by many contemporary writers, the *Burra Record* informing readers in 1891 of the sad case of Samuel Curgenwin who had left the Burra to mine at Broken Hill, but was now "leaded" and confined permanently to bed. In the following year the same newspaper wrote that,

Big stalwart men who left the Burra and Moonta Mines just a year or two ago to toil in the silver mines of the Barrier, are now in many instances past recovery, and the rest of their days apparently must be spent in helpless misery.

Similarly, in 1904 Donald Clarke wrote that,

In no town in Australia can one see so many men propped up against walls, or aimlessly wandering about; the women age rapidly, and even the young children have old faces. Fowls, which picked up the surface soil, and cats, who cleansed their fur, soon succumbed, while many children were leaded in this unhealthy town. Miners, working amongst the dust of carbonate ores, become leaded, and even now there are many human wrecks left as relics of the boom days of Broken Hill.

And unlike the South Australian copper mines, there were no

company-sponsored welfare schemes to which the disabled miner could turn for help or compensation. Indeed, the old Cornish paternalism - so noticable on the copper fields - was largely absent in the big Broken Hill mines, many of the American managers having little knowledge of Cornish ways, and the Cousin Jacks feeling no ethnic or religious affinity with their "Yankie" bosses.

It was against the background of such conditions that the strike of 1892 occurred, the actual precipitating factor being the renewed attempts by the large mines to introduce the contract system. The unionists opposed contracting because it encouraged dangerous practices, penalized elderly miners who could not work fast, endangered the employment of slow workers, and - of course - because it tended to undermine the solidarity of the miners. There were many ugly scenes during the strike, with troopers being brought in from Sydney, and "free labour" engaged from outside to break the strike. By the November the strike had indeed been broken, the miners defeated and their leaders gaoled, but of particular interest from the Cornish point of view was the support shown for the unionists by the Cornish community in South Australia. David Morley Charleston, the Cornish engineer and Labor leader, travelled up from Adelaide to address the strike meetings, while the Moonta *People's Weekly* newspaper detected a "... deliberately organised, conspiracy against Labor, and first and foremost against Unionism as its strongest citadel." And when E. Polkinghorne, John Bennett, and other AMA leaders were arrested and convicted for conspiracy, Cornish born R.J. Daddow - the Primitive Methodist minister at Kooringa (Burra) - declared that the "martyrs" had shown "... self-sacrifice, self-control and moral courage." Most Cousin Jacks would have agreed with him.

The defeat of 1892 badly damaged the trade union movement on the Barrier. Contracting was indeed implemented at the mines, and output doubled as a result. But profits fell in 1893, wages were reduced, and hours made longer. And although the union was now weak, seeds of bitterness had been sown that in the fullness of time would grow and ripen into damaging industrial conflict. By the turn of the century the trade unionists had regained their self-confidence and their authority amongst the working people. In 1909 wage-cuts following a profits slump led to a 20-week lock-out, and in 1916 there was an especially bitter eight-week strike. Three years later, in 1919, came the famous strike of the 600 days - when the victorious trade unionists won the 35-hour week, compensation of plumbism, and a new, high minimum wage.

By now, of course, the moderate, liberal unionism of the Cornish was hardly relevant. The left-wing International Workers of the World movement had gained considerable influence on the Barrier, as had the Australian Communist Party, and even the Australian Labor Party was quite different from the old United Labor Party that the Cousin Jacks had known in South Australia - Socialism replacing Methodism as the major trait and motivation.

170

Industrial saboutage and arson, almost unknown in Moonta and in Cornwall (the only major exceptions in the latter being at Pendeen Consols and North Pool in the 1850s, and at Great Wheal Busy in 1866), also became a feature of industrial conflict in the "Silver City," an indication of the extent of the bitterness and antagonisms.

Needless to say, the Cornish influence did not disappear entirely overnight - in 1906, for example, the Mayor of Broken Hill was a local Labor leader called Thomas Ivey, a miner from Gwennap and Kapunda. Similarly, Josiah Thomas, one of the early Cornish unionists, entered the Sydney Parliament as a member for Broken Hill in the 1890s and remained there for some years, though later replaced by more radical types. But the general trend was towards the more modern, militant unionism, and as it developed so the trade union links with Yorke Peninsula weakened. To some extent the growing confidence of the Peninsula activists was due to the influence of their Barrier colleagues, but by the turn of the century most commentators were pointing to what they saw as the growing differences between Peninsula and Barrier unionism. In 1908, for example, Ralph Stokes felt it important to compare the "fallacious precepts" of the Broken Hill miners with the moderation of the men at Wallaroo and Moonta, where "... labour difficulties are only experienced in times of abnormal activity, and are speedily dispelled."

In the same way, the emotional, cultural and social links between the two districts began to weaken, although it was a more lengthy process. In the 1890s and early 1900s the connections were still quite intimate. In December 1893 for example the Broken Hill correspondent in the Moonta *People's Weekly* took pleasure in explaining that "... a very large proportion of our population have connections in Moonta, Kadina, and Wallaroo...," while gossip columns in the Peninsula press were full of such items as "Jack's return from Broken Hill is looked forward to with great expectation by Jane." During the festive season there was considerable physical movement between the two communities, one report in the 1890s recording that,

A frequent question just now among ex-Peninsularites is, "Har ee goin ome Xmas?" I believe there is no other town in Australia from which there is a greater exodus of holiday-makers at this season of the year than at Broken Hill, and on this occasion there is quite the usual number leaving for the well-remembered sights of Kadina, Moonta and Wallaroo.

A further expression of these links was a poem published in the *People's Weekly* in June 1893, an account of a romance between a Moonta girl and a Barrier miner boy. The poem begins by recounting how they met and fell in love:

171

When first I was courted by a Barrier miner boy,
He called me his jewel, his heart's delight and joy,
It was in this Silver City, our town of noted fame,
Where this 11 Block mining lad acourting with me came.

The piece continues with a description of the many virtues of the miner boy, and of the couple's happiness, but goes on at length to tell how the miner quit the district, deserting his Moonta girl and leaving her broken-hearted. The final stanza tells its own dismal story:

And when I'm dead and gone, this one request I crave,
You'll take my bones to Moonta and lay them in the grave.
Some words write on my tombstone to tell the passer-by,
I died all broken-hearted through that 11 Block miner boy.

Mining disasters also served to emphasise the links between the two communities, and to draw them together. The Peninsula mines were relatively safe, but - as already noted - their Barrier counterparts were rather more dangerous. When John Olds was killed by a rock fall in the Proprietary Mine in 1895 his body was taken back to Moonta for burial, and the death of Phillip Eddy (who "fell away" in a shaft in 1896) was widely reported in the Peninsula press. Thomas Ninnes, from Cross Roads (Moonta), was killed in the South Mine in 1902, an event which was especially tragic for his mother, who had lost her husband more than 20 years earlier in a similar accident at the St. Ives Consols mine in Cornwall. Even more traumatic, however, were the South Mine disasters of 1895 and 1901. In the former, amongst the dead were two Moonta boys, Arthur Trembath and John Slee (born in Bodmin in 1854), while in 1901 the six men killed included Henry Downs from Yelta (Moonta), John Prideaux from Kadina, and William Bennetta from Wallaroo Mines.

The links between Broken Hill and northern Yorke Peninsula remained important for many years, but by the early 1900s - with the onset of better times at Wallaroo and Moonta - the exodus to the Barrier was stemmed, and thus the Peninsula influence there lost its immediacy and on-going character. Few Cornishmen made their way to the Barrier from other parts of Australia, or from Cornwall itself, and - although still unique - Broken Hill developed in this century as a distinctly Australian town. The Cornish inheritance was just one element of a rich mining heritage, and it was not many years before the term "Cousin Jack" came to acquire a slightly derogatory conotation - to be called "Cousin Jack" meant that one was thought rather old-fashioned, strait-laced, a "country bumpkin," a relic of an earlier era. Today, John Gough, curator of the Broken Hill Museum, adds somewhat whimsically "Yes, everyone knows the Cousin Jacks were important, but not a lot has survived, nothing much has been written down." The real memorials to the Cornish are not in words, but are physical things -

172

the slag-heaps along the "line of the lode," the ripple-iron homes and especially the old chapels.

CHAPTER 7

THE GOLDEN WEST

The 1880s were notable for the widespread movement of Cornish miners from South Australia to the Barrier district of New South Wales, but the following decade witnessed a similar large-scale migration of Cousin Jacks from the Central Colony (and to some extent, too, from Broken Hill) to the newly-discovered gold-fields of Western Australia. Indeed, the Westralian Gold Rush of the 1890s gave the continent's "western third" not only its first major influx of Cornishmen, but also its first real opportunity to develop a mining industry and provide an economic base to support substantial population growth. Before then, Western Australia (or Swan River Colony as it was known at its foundation in 1829) had been largely an uninhabited wilderness, its 48,000 people in the early 1890s being concentrated in the relatively fertile south-western corner of the colony, many in the capital of Perth itself. The other Australian colonies, with their larger populations, prosperous industries, and significant mining fields, were - to quote one contemporary source - "... all apt to look upon their western sister with a tolerant, patronising air."

There had, though, been some limited prospecting and mining in Western Australia before the boom of the 1890s. As early as March 1852 the *West Briton* had carried advertisements encouraging the emigration of Cornishmen to Swan River Colony, one entry that month calling for "... steady, active, young men having a thorough knowledge of copper mining." These recruits were presumably required for the modest copper workings (the Geraldine mines) then being opened-up between Geraldton and the Murchison River, on the western coast nearly 300 miles north of Perth. Lead was also found in the area, and in 1867 Captain Samuel Mitchell arrived from Cornwall (where he had been born in 1839) to manage the Geraldine Lead Mine, near the small township of Northampton. Mitchell also became manager of the Badra lead workings, some seven miles from Northampton, and was responsible for the opening-up of Wheal Ellen - another silver-lead mine in the area (apparently named after South Australia's Wheal Ellen at Strathalbyn), which yielded some £70,000 worth of metal. Although never spectacularly rich, the Murchison-Geraldton field afforded intermittent employment for many years, James Whitburn from West Cornwall arriving there as late as 1874 to work in the Old Geraldine copper mine. Whitburn later made his way to the copper workings at Great Cobar in New South Wales, and from there to Broken Hill, returning finally to Western Australia during the Rush of the 1890s.

Other prospectors and miners had been attracted to Western Australia by the chance that there might be hitherto undiscovered gold bonanzas in the little known outback areas. In 1873 a party of 16 Ballarat miners was brought in to look for gold on behalf of one local concern, but was apparently unsuccessful. By the early 1880s, however, goldseekers who had made their way up through Queensland, from New South Wales and Victoria, to the Northern Territory were beginning to cross the border into the barren north-western wilderness of Western Australia. Two diggers with Cornish names, Captain Adam Johns and Philip Saunders, were amongst the first. They set out from the Pine Creek mine in the Territory, and in August 1882 began to explore the arid ranges known as the Kimberley region. They found some gold in the creek beds, but they could never find enought water to wash the sand efficiently and systematically. They were harrassed by hostile Aborigines, found many of the ranges almost impassable, and were constantly racked by thirst and sandy-blight. Johns on his return could barely walk or see, but - like others in the Territory whom their expedition had inspired - they later went back to Western Australia in their quest for gold. Saunders remained a digger all his days, until his death at the great age of 93 when he fell into a camp-fire in the goldfields town of Menzies.

The path-finding work of Johns, Saunders, and the other diggers from the Northern Territory, prepared the way for an influx of miners from other parts of the continent, and by May 1886 the Kimberley Rush was in full swing. Some Cousin Jacks travelled up from Moonta, and others from the goldfields of Victoria taught the techniques of "dry-blowing" - how to extract gold from sand without the use of water. Although the dust it caused was unpleasant, the method was simple - a pan full of sand was held aloft and tilted, the light material blowing away while the heavier stuffs (including any gold) fell into another pan placed on the ground. Disease and heat, not to mention the indifferent nature of the fields, killed the Kimberley Rush, however. Some, like St. Just-born Captain James Newton (a former underground manager of the Levant mine in Cornwall), found their "final resting-place" in the ranges, while by 1888 many of the diggers were moving southwards across the Great Sandy Desert to the Pilbara region. Known today for its rich iron mines, the Pilbara was then explored on account of the modest gold it had to offer, and although its rush was hardly successful it did have the effect - as had the Kimberley Rush - of drawing experienced gold-seekers to Western Australia for the first time, many of whom were to play a significant role in the great 1890s Rush. There was, for example, Captain Thomas Gilbert Pearce, a man born in 1842 "... within a mile of the famous St. Michael's Mount..." (probably Marazion), who had emigrated to Victoria in 1857 and had dug gold in that colony and New Zealand for many years before moving on to the Marble Hill area of the Pilbara. Later, Pearce was to become an important manager and entrepreneur on the Coolgardie fields.

And while the Kimberley and Pilbara diggers continued to push south, other goldseekers - inspired by the limited success in the north - moved eastwards from Perth, into the outback. In 1887 the Yilgarn goldfields were discovered, and in 1888 there were further finds at neighbouring Southern Cross, almost 200 miles east of Perth. Amongst the early diggers at Southern Cross was Captain William Oates, who was born at St. Just-in-Penwith in 1842. Oates, a former captain at Wheal Owles (a mine on the cliffs near St. Just), was persuaded to migrate to Victoria in 1884 by George Lansell, the "Bendigo quartz king." He spent only a short time at Bendigo, however, before moving to the Barrier to become captain of one of the Euriowie tin mines. Soon after, he was in Adelaide - and from there he travelled to Western Australia to work the Fraser South Mine at Southern Cross. It was there, so it is said, that Captain Oates became the first man in the West to smelt gold. Certainly, his reputation spread quickly and, with the dawning of the great Rush in 1893, his services as consulting engineer were soon in demand by many of the colony's gold companies. In May 1897 the *Kalgoorlie Western Argus* expressed the opinion (held by many on the goldfields) that "Of all West Australian mining engineers, Captain Oates is *the* mining engineer par excellence."

Taking heart from the success of Southern Cross, bands of diggers continued to press eastwards into the semi-desert. And in June of 1892 Arthur Bayley and William Ford, two prospectors with a long record of successful digging in nearly all the Australian colonies, discovered a spectacularly rich gold deposit at an isolated location known as Coolgardie, some 350 miles from Perth. With news of the find spreading fast across the colony, diggers hurried to the area to mark out their own claims, and by 1893 the Rush was well and truly under way. Paddy Hannan, an Irishman, gave further momentum to the Rush when he discovered a second and equally spectacular deposit at neighbouring Kalgoorlie.

By now, news of the Rush had reached every mining camp in Australia. In Adelaide, the mining capital of the continent, numerous syndicates were formed and companies floated. One such concern, the Adelaide Prospecting Syndicate, hired the services of two local well-known prospectors, W.G. Brookman and Sam Pearce. Pearce hailed from the copper town of Kapunda, and had mined in California, Mexico and South Africa, as well as in most of Australia. In the 1880s he had been on the Rand, but had returned to search for gold in the Adelaide Hills. But there was scant gold to be had in the Hills, and Pearce would have needed little persuading to join the Syndicate's band. He and Brookman set sail from Port Adelaide on 7th June 1893 and landed at Albany, whence they travelled overland to Perth and then Coolgardie. They arrived in the goldfields on 29th June, but hearing of Hannan's find they decided to go on to Kalgoorlie. Disappointment greeted them there, however, for they found almost all of the ground around Hannan's claim already pegged out. Thus they moved on to the nearby Ivanhoe Hill which, although already dismissed as worthless by most diggers, gave Pearce and Brookman and lucky find they so badly needed. Indeed,

Great Boulder (the name they gave to their mine) proved as an important a find as Coolgardie and Kalgoorlie, and Sam Pearce was able to peg-out many other important claims in the area - Lake View, Three Australias, Iron Duke, Royal Mint, Consols, Associated.

In the wake of the discoveries at Coolgardie, Kalgoorlie and Boulder, further bands of prospectors pushed even further into the outback, to districts with names as apt and forbidding as Siberia and Dead Finish, where water was virtually non-existant and death awaited many a luckless digger. James Tregurtha was one Cousin Jack who tackled the far-outback area. He and his mate, Billy Frost, went first to Siberia, and from there made the amazing trek across the worst country in the continent to the opal-mining town of Oodnadatta in northern South Australia. On their return journey to the West, they lost their horses and Tregurtha and Frost were obliged to complete their mammoth journey on foot!

Men like Tregurtha were the exception, however. Most newcomers to the Westralian goldfields were content to become employees of the large mining companies which sprang up after 1894. Indeed, the transformation from small-claim prospecting to deep, heavily-capitalised mining was remarkably fast - far quicker than that experienced in earlier years in Victoria, and even faster than at Broken Hill. And associated with this rapid expansion was an ever-increasing influx of migrants from other parts of the continent - 16,000 arriving in 1894, 18,000 in 1895, and 36,000 in 1896. Not surprisingly, many were from the adjoining colony of South Australia (as late as 1933 the Census showed more South Australian-born males than Western Australian-born), a considerable proportion of whom were Cornish or of Cornish descent.

By April 1894 the *Register* newspaper, published in Adelaide, was noting the departure of Yorke Peninsula miners for the West, the Moonta *People's Weekly* in the same month recording that Richard Ellis of East Moonta and his three mates - messrs. Thomas, Arthur and Pryor - had struck a rich gold reef near Coolgardie. In the August one observer could write that,

... the Western Australian Goldfields ... are attracting large numbers of miners and mechanics from both mines (i.e. Moonta and Wallaroo). I am pleased to find that the West is an outlet for miners - and more especially for those of Yorke's Peninsula, who are preferred above any other in the colonies.

Bands of unemployed miners continued to leave Yorke Peninsula during 1895 (the mines had yet to recover from the depression of the 1880s) in a fashion reminiscent -as one commentator put it - of "... the palmy days of Broken Hill." Yet another observer expressed the view that "... this exodus... will prove as successful in taking away the surplus labor from this district as Broken Hill did...," while one Western Australian wrote that on the goldfields

177

"... your miserably ill-paid Moonta miners (have) a chance to earn a decent livelihood." There were further departures from Wallaroo and Moonta during 1896 and 1897, and another wave left the Peninsula in 1899. Others - such as C.R. Treloar, H. Martin, and Walter Rosewall - left the Burra to go to the West, and there was also a sprinkling of Cousin Jacks from other parts of the colony - Richard Alford from the Adelaide suburb of Bowden, Henry Bastian from the northern settlement of Booleroo, William Martin from the rural township of Willowie, and Thomas Harvey from Prospect (Adelaide) who was killed near Kalgoorlie when his tent was struck by lightening.

Some made their way to the goldfields by sailing from Port Adelaide to Fremantle (near Perth) or Albany, and then travelling inland, but many others chose the overland route. It was some 1200 miles from northern Yorke Peninsula to Coolgardie and Kalgoorlie, much of the journey through sandy, scrubby wilderness. The series of letters written by Daniel and David Williams to their parents while travelling the overland route from Wallaroo affords many insights into conditions on the route. They left the Peninsula in the early part of April 1894 and by May 30th had reached Eucla, a border town set in the heart of the arid Nullarbor Plain. There they were obliged to pay £8 10s 0d customs duty on their horses (hopes for inter-colonial free trade, let alone federation, were then only dreams!), writing that:

... we had a pretty hard time of it (or rather the horses did) coming from the Bight here the last 3 days we couldn't get any water for them for there was teams in front of us and they took the lot but we got through all right...

By mid-August the Williams brothers were nearing the goldfields. They wrote from Diamond Rock Station, explaining that there were many Wallaroo men heading for the West and that,

... there are a terrible lot of teams coming the overland trip and some of them are getting a rough time of it, there's all sorts travelling. Waggons, vans, spring drays, buggies, pack horses, camels, and foot men. Some have come from Queensland, New South Wales and Victoria.

Daniel and David began prospecting during the September. At first they were intensely optimistic, writing that "Western Australia is a great country for gold...," but by early November they were beginning to realise how difficult it was to "strike it rich" and that life on the goldfields was hardly rosy. David informed his parents that,

... I can't write anything interesting about our trip, since we got into the colony its been nothing else but hunting for water. I

hope we will get rewarded before long... I don't think this place
or these fields are what they are cracked up to be.

Daniel wrote again from Coolgardie on 26th November noting that,
with the approach of summer, "... there is a good bit of sickness here now,
there is not enough nourishment in the tucker ..." His letter of 10th February
1895 showed him to be in low spirits - "... we spent our Xmas in a lonely spot
where we could see nothing but scrub... I must get a little gold somehow..." -
and both Williams brothers realised that there was little success to be had in
prospecting as individuals. Daniel, therefore, found work as a miner in the
large Bendigo Coolgardie venture, while David was fortunate enough to
acquire a small wheelwright's business in Coolgardie township. With this
change in their luck, Daniel believed that "... there's a grand future for W.A.
There's a lot of good mines in this country..." And because "... there's a lot of
females getting here now, and the place is getting quite toffy, Dave and I are
beginning to wish we brought our Sunday clothes with us." This mood of
renewed optimism did not last long, however, for the arrival of hot weather in
early November was a reminder of Coolgardie's unpleasant summer
conditions. The brothers wrote that, "There will be a lot of sickness here this
summer ... water is getting scarce here now, they are charging 6d a gallon.. it's
no use stopping here, we might get sick..." They decided, therefore, to quit the
goldfields and make their way to Perth to seek employment there, and
thereafter - their goldfields sojourn over - their correspondence ceases.

Perhaps Daniel and David were wise to leave when they did, for
unpleasant social conditions - heat, fever, periods of unemployment, and so
on - were a subject mentioned in almost every letter home (wherever "home"
might be). During January the average temperature at Kalgoorlie was 100° F,
and the Cornish miners spoke with their typical dry humour of "Hotgardie."
One observer had this to say to Kalgoorlie:

Mulga, quandongs, cotton-bush and salt-bush hide the
bareness of the parched red soil, while on the sand ridges the
globular tufts of the bright-green spinifex shows in marked
contrast against the blue of the short scrub and the bilious
green of the scraggy gums. Great flat plains, covered in some
instances with sand, in others with stunted scrub, are known as
lakes (sic)... The soil appears to be infertile, but is redeemed
from being a desert by the scrub, which extends over hundreds
of miles. As soon as settlement destroys vegetation the fine red
soil which floats with every breeze envelopes and tinges all
natural and artificial objects with a color peculiar to itself.

As early as September 1894 "Nine Moonta Boys" wrote back to Yorke
Peninsula reporting water shortages and unemployment, while in the

November a correspondent to the *Cornish Telegraph* newspaper considered that,

Western Australia, taken all round, is the most God-forsaken place a man can set foot in. It is an awfully rough life here. I have not slept in a bed for over three months, and only get a wash once a fortnight, the food is also bad; no fresh food of any description - all tinned; and now the weather is very hot it is all in a liquid state... You can't buy water even in this great city of Coolgardie under nine-pence per gallon. I know a man who paid five pounds for watering his three horses for one night. We have been to a good many "new" rushes with no luck at all. At one of them the discoverer of a reef had pegged the land for miles, and when we arrived water was five shillings a gallon. There is no sport of any kind out here - not a living animal except flies, ants, and snakes.

In November of the following year the Moonta *People's Weekly* recorded a new wave of disease and water famine, while in January 1896 an East Moonta man wrote that "... Coolgardie is alive with unemployed..." and that water was 10d a gallon, flour 6d per lb, bread 11d per loaf, and jam at 1s 6d a jar. Twelve months later the situation had changed little, one Peninsula miner at Boulder describing the goldfields as "... the land of swags, rags and water bags..." It was not unusual to see death notices such as the following in the *Kalgoorlie Western Argus* - "A young man named John Cocking has died in hospital from fever. He was a native of Cornwall" - and the behaviour of some of the miners only served to aggravate the conditions. In March 1895 three diggers from Kadina in South Australia were under arrest for "... the wholesale murder of blacks...," and in March 1896 a Cousin Jack named Bluett was in trouble for shooting off his half-brother's finger, while another Cornishman (called Joseph Rundle) was fined £3 for running an illegal gambling den. But it was disease which caused the most distress and claimed the most lives, it being said of one Joseph Lathlean in 1898 that,

Like many others from this district (Moonta), owing to the low rate of wages on the Peninsula, he was forced to seek employment elsewhere; hence he went to Western Australia, where he was stricken with pneumonia which left a weakness which soon developed into consumption, and gradually he faded from us.

And disease was no respector of person or status, one goldfields newspaper in December 1897 noting the death at Coolgardie from dysentry of Captain Charles Thomas Rowe. Captain Rowe, manager of the Northern Wealth of Nations and Central Wealth Consolidated Goldfields mines, was a man of some standing in the community, for - in addition to his considerable experience in Cornwall and Australia - had mined in Corsica, India and Ireland. Mining disasters were no respector of rank either, the *West Briton* in March 1897 recording the death at Coolgardie of Captain John Brokenshire from Roche. Brokenshire had been inspecting a winze in a local mine when killed suddenly by a premature explosion. Then 66 years of age, he had emigrated to Australia in 1893 with his colleague Thomas Parkyn - "... the well known explorer and mining expert..." from Roche, after a varied career in Britain and South America.

A number of miners from northern Yorke Peninsula lived in the so-called "Moonta Camp" at Boulder, which, judging from the following composition of December 1896, was a decidedly unpleasant and unhealthy locality:

'Tis a number of camps on rising ground,
And a few lateens scattered around
    Of the same style of architecture:
They're built of poles and old chaff bags,
Canvass, calico tents, and rags,
    All of different hues and texture.

Tins and bones are lying around,
    Bags and other filth abound,
      And things of a similar stamp;
And fever germs have a depot there,
    And a horrid perfume fills the air,
In the Boulder Moonta Camp.

The Moonta Camp looks old and scarred,
The hill they're on looks strange and wierd -
    It oppresses one with dread;
And the costeen pit, where the lateens wave,
Looks like a huge uncanny grave
    Awaiting for the dead.

The existence of a Moonta Camp (reminiscent of the Moonta Town at Broken Hill in the 1880s) indicates that the Peninsula folk retained the Cornish tendency to "stick together" when amongst "foreigners" and away from home. The survival and use of the Cornish word "costeen" (a miner's exploratory trench) in the poem is also of interest, but what is conveyed most of all is the sense of destitution and squalor.

But, however bad conditions in Western Australia may have been, this did not deter the more ardent bands of goldseekers from travelling from

181

northern Yorke Peninsula (and, indeed, many other localities) throughout the 1890s. The very existence of the Boulder Moonta Camp was evidence of the large numbers of Peninsula Cornish in the West, while one observer could note that "... Moonta Camp represents only a very small selection of Moonta boys on the field..." It was certainly true that the South Australian Cornish could be found throughout the Coolgardie-Kalgoorlie-Boulder district, and on "neighbouring" goldfields as far away as Menzies and Norseman, and it was never long before new arrivals from the Peninsula came across friends from home.

At Coolgardie in December 1894, for example, Daniel and David Williams from Wallaroo (the two brothers who had trekked across the Nullarbor) met "... several Wallaroo chaps..." - Harry James, Tom Smith, Jim Samuels, Joe Phillips - while in August 1895 they wrote that "There's a lot of Moonta and Broken Hill people coming here lately." In March 1896 there was "... a lot of Wallaroo's over this way...," while Stanley Whitford from Moonta Mines wrote that on the goldfields at the turn of the century "... I was camped with a nest of Cousin Jacks from Moonta." There were, he said, Jack Pascoe, Stan Verran, Alf Northey, Jos Liddicoat, and Merts Trebilcock. Bill Roberts and Jack Warwick, both Moonta men, were underground captains at the Australian and Great Boulder mines, while Tom Horton - then manager of the Malcom Proprietary - was a former captain of the Yelta copper mine, near Moonta. The existence of a syndicate named the Yorke's Peninsula Gold-Mining and Prospecting Co. Ltd. was still further evidence of a Peninsula influence, and Moonta men - such as those who discovered gold assaying at 100 ounces per ton at Kalgoorlie in September 1895 - were amongst the small minority of diggers who really did "strike it rich."

In January 1897 it was said that there were "... more Barrier men in Kalgoorlie than ... in Broken Hill..." Certainly a considerable number of the Westralian gold mines were managed by the South Australian and Barrier Cornish. In January 1895 it was reported that the Royal Mint, Lake View, Great Boulder, Australian, and Iron Duke mines were run by men "... well known to Kadina and Broken Hillites...," while the captain of the Ivanhoe was "... an old Moonta identity...," and W. Rowe - manager of the Maritana and Napier claims - was "... late of Moonta." The surveyor at the Great Boulder was also "... an old Moontaite...," while the mine's administrative clerk was one M. Rodda - formerly an accountant at the Block 14 at Broken Hill. At the Faith mine the captain was Charles Truscott, an experienced Cornishman who had worked in Wallaroo Mines and at Ediacara in the far-north of South Australia, and the Associated claims were run by James Morton and Charles Davey - both from Moonta.

In June 1895 B. Nankivell, "... well-known to Moonta and Kadinaites...," was appointed Chief Captain of the Great Boulder, while in the December Captain East from Adelaide was at the Hannan's Central. Even

Moonta's Captain Hancock got in on the act, successfully recommending his son Leigh for the manager's job at the Boulder Central. Another Moonta man, Tom Warren, was appointed captain of the Great Coolgardie, while W.F. Francis (also from Moonta) was at the Adelaide Sovereign and Kalgoorlie Reef mines. Alex Roberts, yet another Peninsula miner, was captain at the Kanowna Consolidated, and the Adelaide-owned Arrow Proprietary was in the charge of one Captain William Hambly. Other Cornish managers included William Begelhole at the Bayley's Reward, Frederick Rodda (the "... popular and efficient..." captain of the Hit and Miss), Thomas Pascoe (Vice President of the Kalgoorlie Mine Managers' Institute in 1897) at the True Blue, and Captain Dunstan at the Hannan's Consols and Colonial Goldfields mines. On occasions Cornishmen came from far further afield than Moonta or the Barrier, as in January 1897 when the *Kalgoorlie Western Argus* reported that:

Mr. William J. Barnett, a medallist of the Mining Institute of Cornwall, has arrived as the representative of the Mines Selection Company, Limited. He is a well-known mining engineer, and has been appointed to succeed the late Captain Vaundry, whose death was so generally regretted. Mr. Barnett was specially recalled from South Africa to undertake the duties entrusted to him.

Although the Cornish were prominent as mine managers, only a relative few emerged as major capitalists and directors of the large concerns. One exception was Captain Thomas Gilbert Pearce, who had arrived in the colony in 1893 to dig at Marble Hill, moving on to Coolgardie in 1895 where he floated a string of successful gold mines - the Richmond Gem, the Irish Lily, the Lady Loch, and the Easter Gift. But perhaps the most celebrated Cornish personality in the West was Helston-born John Treloar. Treloar had started out as a Burra miner but had mined in Victoria, on the Barrier and in New Zealand before making a small fortune as an entrepreneur on the Westralian goldfields. He was one of the original owners of the Brown Hill, a working which was amongst the most successful of the early gold mines, and visited London to promote investment in the colony. One contemporary account recorded his progress:

Captain John Treloar, well-known to most Australian mining men, has been on a visit to Cornwall - his birthplace - where he has been giving some attention to Cornish mining. He returned to London on Monday (in July 1896) and he will leave for Australia at the beginning of August... Captain Treloar, who has had 47 years' practical experience in Australia and New Zealand, and who has been associated with some of the best

183

Westralia finds, notably McAuliff's Reward, Reefer's Eureka, and Hannan's Brown Hill, has done a certain amount of business here (London), but he thinks that he might have done far better had he taken his Westralian properties to Adelaide. On arrival at Albany, Captain Treloar will proceed to Coolgardie to report progress. He will then return to Adelaide to rejoin his wife.

Although there were few Cornish capitalists of Treloar's calibre, Adelaide investors displayed a keen interest in Western Australian mines, and the Wallaroo and Moonta Mines company made alterations to its smelting works so that it could smelt gold from the West. J.J. East (a son of Captain East from Calstock) was appointed the company's agent in the West, and the Directors decided that, in purchasing gold for their smelters, "... the Cornish Ticketing System is the most suitable for the Mines having Gold ores to sell." Some Westralian mine names also revealed a Cornish influence - there was a Devon Consols (named after Devon Great Consols on the banks of the Tamar), and there was of course Wheal Ellen silver-lead mine.

The real Cornish influence was at the "grass roots" level, however, in the behaviour, attitudes and activities of the ordinary Cousin Jack miners. At the Moonta Camp the men would spend their evenings playing euchre, and at Christmas could be heard singing their traditional Cornish carols. They also formed a Moonta Camp cricket team, and at Boulder the local footballers were known as the "Amber-and-Blacks" on account of the traditional Cornish colours that they wore. The Cornish, with their great love of music, also featured prominently in bodies such as the Boulder Choral Society and the Boulder Mines band. And, of course, they had their wrestling - in September 1895, for example, there was a much publicised contest between P. Roachock (sic) of Norwood (Adelaide) and Harry Pearce of Moonta.

Items of "Moonta News" appeared in the goldfields newspapers from time to time, as did reports of happenings in Cornwall. In the February 4th 1897 issue of the *Kalgoorlie Western Argus,* for instance, we learn that Mr. George Hannaford of Millbrook has been committed for trial for the murder of his father, following an inquest at Torpoint. And, not surprisingly, considerable space was devoted in March of that year to details of the world heavy-weight boxing championship when Cornish-born Bob Fitzsimmons defeated "Gentleman Jim" Corbett in America.

There were countless other manifestations of emotional links with Yorke Peninsula and Cornwall. When one Boulder miner named his racing pigeon "Moonta Lad" he was thinking fondly of home, as were the diggers who in 1896 contributed to the memorial fund for Arthur Trembath - the Moonta boy killed in the South Mine accident at Broken Hill. When John Treloar was in Britain on business he made a point of visiting Cornwall, as did Frank E. Allum, a metallurgist who was to become Deputy Master of the

184

Royal Mint in Perth, Western Australia, in 1928. When in Cornwall Allum visited the Levant mine, near St. Just, writing a rather turgid and pedestrian account of the event which was reprinted in the *Old Cornwall* magazine some years ago.

As elsewhere in Australia, the survivals of Cornish culture in the West included the usual sprinkling of Cornish sayings and superstitions. There were clearly memories of the seventeenth-century rhyme "Hingston Down well-y-wrought, Is London town dear-y-bought", for an up-dated version - "Caradon Hill well wrought, Is worth London Town dear bought" - was current on the goldfields. Perhaps not so exclusively Cornish, but of the same genre, was the goldfields "marriage rhyme:"

> Married in white you have chosen all right.
> Married in grey, you will go far away.
> Married in black, you will wish yourself back.
> Married in red, you will wish yourself dead.
> Married in green, ashamed to be seen.
> Married in pearl, you will live in a whirl.
> Married in yellow, ashamed of your fellow.
> Married in brown, you will live out of town.
> Married in pink, your spirit will sink.

The first wedding at Coolgardie, as it happened, was between an apparently Cornish couple - Miss Clara Saunders and Mr. Arthur Williams - and the officiating clergyman was the Rev. Thomas Trestrail, a Cornish Methodist who had already spent some years working amongst the miners at Broken Hill. His colleague, the Rev. G.E. Rowe, became the first President of the Western Australian Methodist Conference and was described by the *Australian Christian Commonwealth* magazine as a man "... who can tell a Cornish story exceptionally well..." Later, as the *Cornishman* newspaper noted in 1902, Rowe returned to Cornwall, where he took up duties in the St. Austell Circuit.

Needless to say, Cornish Methodism was an important influence on the goldfields. At Boulder there were three Methodist chapels, each denomination with its committed members "... many of them Cousin Jacks..." - and the Orange Lodges were strong. On 12th July 1896, for example, it was noted that 100 Orangemen marched through Kalgoorlie in the Battle of the Boyne parade, "... several Moontaites being conspicuous." But despite their religious differences, the Cornish and the Irish did try sometimes to get along together, as in March 1897 when Cousin Jacks with names such as Moyle and Bawden were reported leading the St. Patrick's Day celebrations.

Like the eighteenth-century Cornish smugglers (the "free traders"), the Cornish miners had occasionally somewhat liberal interpretations of Methodist morality, the tradition of "making a bit extra" insisting that God intended a proportion of the gold for the miners themselves. Casey and

Mayman, in their charming book *The Mile that Midas Touched,* re-tell an old Cousin Jack yarn about a "... Cornishman... practising a hymn for the church choir as he picked over the stone. 'Do Not Pass Me By,' he sang, and softly the words echoed through the dark stope as put aside what he looked on as his portion of gold..."

This tradition of "making a bit extra" was sometimes a cause of conflict however, as in March 1896 when William Berryman was charged by Captain Pollard with stealing over £15 worth of gold from Bayley's No. 1 South mine. And, although industrial conflict was never as marked as at Broken Hill - or even South Australia - a vigorous trade union movement did emerge. The Cornish wrestled with the Irish for control of the goldfields union (the Kalgoorlie and Boulder Branch of the Amalgamated Miners' Association was formed in August 1896, and a combined goldfields executive set-up in 1897), but Professor Geoffrey Bolton has noted that "Because so many South Australians came to W.A. in the 'nineties gold-rush they had a considerable impact on the character of the early Labor movement here."

Bolton also argues that a number of Western Australian politicians, including several colonial and State Premiers, were "... products of the Cornish-South Australian tradition." It was perhaps a measure of this influence that contracting (much abhorred by Moonta and Wallaroo men) was not introduced in the major mines until the 1920s, and the impact can also be traced in both public attitudes and Parliamentary legislation. The most colourful personality in the early goldfields Labor movement was Captain William Oates, the St. Just - born mining engineer who had first smelted gold in the West. In 1894 he was elected Mayor of Southern Cross, and in 1897 he stood as a Labor candidate for the Parliamentary seat of Yilgarn, on the goldfields. A contemporary account recorded that,

He has risen from the ranks, and he believes in men rising from the ranks. He believes in a mutual ground of co-operation between employee and employer. He would like every employee to own a share in the mine in which he is working.

Oates, therefore, was an early advocate of "industrial democracy," and it was certainly a moderate form of socialism that he was pressing for. He enjoyed widespread support amongst the miners, and it was no surprise when his attempt to enter Parliament was successful. Like many other Labor men of his day, Oates believed that immigration threatened the jobs and living standards of the workers, and he was a fervent adherent to the "White Australia" policy, one report declaring "... William Oates has never employed a nigger in his life, and never will."

One of Oates' Parliamentary colleagues was Captain Samuel Mitchell, the Cornishman who had migrated to Western Australia in 1867 to mine near Geraldton, a man known for his "... liberal and progressive activity..." who was returned unopposed to the colony's Legislative Assembly in 1897. Another member, elected in the same year, was one F.C.B. Vosper, a "... liberal and democratic..." politician from St. Dominick in East Cornwall (where he was born in March 1867). In Queensland Vosper had been imprisoned for his part in a miners' strike at Charters Towers, moving to Western Australia in 1892 where he became editor of the *Coolgardie Miner* newspaper, in which he displayed his "... hearty regard for the working people." A striking figure with jet-black hair, Vosper was known (and feared) across the colony. But he died in his early "thirties," before he could really make his mark. Vosper's successor at the *Coolgardie Miner* was Henry Kneebone, a digger from Wallaroo Mines. He, too, was a Labor man, staying in the West until 1910 when he moved to Adelaide to join the staff of the Labor-sponsored *Daily Herald*, becoming editor in 1911 and retaining his post until the paper's collapse in 1924. In 1925 Kneebone entered the Adelaide Parliament, and in 1931 he was elected to the Federal Senate. But despite his success, he never lost sight of his Cornish mining background and was for many years active in the Cornish Association of South Australia.

Stanley R. Whitford was a man especially proud of his Cornish descent, and of his heritage as a miner. Born at Moonta Mines, he was amongst the many who travelled West in the 1890s Rush. And, as he explains in his ernormously long (and as yet unpublished) autobiography, it was on the goldfields that the became a committed Labor man, when elected chairman of a strike committee at the Lake View mine in Christmas 1906. It was there, he said, that "... I developed a realisation of my relative insignificance and the only channel to follow was through the collective effort of the miners..." Whitford later returned to Yorke Peninsula, joining the local miners' union and becoming active in the anti-conscriptionist movement during the Great War. He entered the Adelaide Parliament as a Labor member in 1921, and remained there intermittently until the 1930s.

Like Coolgardie, Kalgoorlie also had its own newspapers - one being the *Kalgoorlie Miner,* a journal founded by the Hockings, a Cornish-South Australian family from Mount Barker, near Adelaide. From the same small township came George Pearce, a man "... of Cornish stock..." who made his way to the Westralian Rush in the 1890s. Peter Heydon, his biographer, tells us that "Though essentially a moderate... he (Pearce) was an agitator and activist...," and by 1910 he was representing Western Australia as a Labor member of the Federal Senate. The conscription issue, however, brought him into conflict with his Labor colleagues, forcing him into the more right-wing National Labor Party, though Heydon insists that "... Pearce was not a bigot. A Protestant and of Cornish extraction, he tried to keep the whole conscription issue free of sectarian and Irish political issues."

Others, though, were keen to point to the alleged Fenian and Popish motives of the anti-conscriptionists, and none was more aggressive than the more extreme Cornish Methodist. In particular, one thinks of Jack Scaddan -

the Labor Premier of Western Australia in 1916 who, like Pearce, resigned from the Labor Party to join the National Labor camp, thus losing his Premiership and ruining his political career. In many ways it was a tragedy, for Scaddan, who was born at Moonta in 1876 and had made his way via Victoria to Coolgardie, had been a model Premier. Fired by liberal and socialist ideals, he had entered the state Parliament as member for the goldfields seat of Ivanhoe in 1904. He became Premier in 1911, addressing himself to such issues as health, education, and the conditions of the goldfields miners.

Another Cornish-South Australian to win the Westralian Premiership was Albert Redvers Hawke, born at Kapunda in 1900, who was member for Burra Burra in the Adelaide Parliament from 1924 to 1927 before moving his political career to Perth. Thus the Cornish influence in the West was projected well into the twentieth-century, although - as at Broken Hill - the real folk-culture of the Cornish miners disappeared quickly as they became assimilated into the distinctive, cosmopolitan, and yet totally Australian character of the goldfields towns. But, as at Broken Hill, the contribution of the Cousin Jacks was not forgotten. When Captain Henry Richard Hancock resigned his position at Moonta and Wallaroo he toured the mining fields of Australasia, remarking in September 1901 that,

He had heard a great deal about them (the Cornish) from those in authority over them, but in no single instance had he ever heard one word against them as miners, but on the other hand he had heard a great deal in their favour. In Kalgoorlie and other places in the West they were in the front rank as miners...

By the early 1900s, however, the flow of miners from Moonta and Wallaroo (the West's major source of Cousin Jacks) had all but ceased, the early twentieth - century witnessing a resurgence of the Peninsula copper mines. By 1915 the Westralian goldmines had paid some £25,500,000 in dividends (much of this coming from the big concerns such as the Great Boulder, the Sons of Gwalia, and the Ivanhoe), but by then the majority of mines outside Kalgoorlie's "Golden Mile" were already in decline. This also helped to stem the flow of miners from outside the State, while the local capitalists began to look elsewhere for workings in which to invest their money. Some dabbled in Mexican silver, but one company - the Hannan's Pty Ltd - sunk its funds into a near-defunct tin mine in Cornwall! Such are the ironies of history.

CHAPTER 8

FAR AND WIDE

South Australia in the 1840s and 1860s, Victoria in the 1850s and 1870s, the Barrier in the "eighties," Western Australia in the "nineties" -at first glance the Cornish involvement in the development of Australian mining appears quite straightforward. But behind the simple chronology is a more complex series of migrations to and fro across the continent, together with an involvement in countless ventures in far-flung corners of the Antipodes.

Tasmania, not surprisingly, features in the story. Inevitably, there were Cornishmen amongst the convicts who formed the earliest population of Tasmania (or Van Diemen's Land as it was first known), but by the 1820s there was a steady - though small - stream of free settlers to the colony from Cornwall. In March 1828, for example, the *West Briton* carried an announcement to the effect that the "Henry Wellesley" - then lying at Plymouth - was due to sail shortly for Van Diemen's Land, and that prospective migrants could obtain details from the Rev. W. Lawry of St. Austell (many clergymen embraced the cause of emigration, another enthusiast was Parson Child of St. Dennis). Ten years later the pace had quickened, and by 1841 a free passage system was in operation.

Issac Latimer and A.B. Duckham, the South Australian migration agents, were also given the task of recruiting for Van Diemen's Land. They advertised in the *West Briton* (as in September and December of 1841), and Latimer issued his usual posters. One such poster, aimed at mechanics, handicraftmen, agricultural labourers, carpenters, quarrymen, masons, and domestic servants, all to be between the ages of 14 and 35, declared that "The Colony of Van Diemen's Land has been established more than half-a-century and possesses the usual advantages belonging to the Australian settlements." In particular, said Latimer, the colony "... is not subject to drought, and affords a peculiar demand for the classes above-named." As ever, Latimer and Duckham were aided in their work by glowing reports from the colony, the *West Briton* reporting in May 1843 that "The ship 'Orleans' which left Plymouth last year with Government emigrants, many of whom were from Cornwall, arrived at Hobart Town (the colonial capital) on 4th July last with all her passengers in good health." Significantly, the report added that "The emigrants met with immediate employment at excellent wages."

But not all observations were so rosy. There were those who pointed out that Latimer himself had already written that South Australia was the superior colony for it had no convicts and did not experience the "... vice and demoralization ... of Van Diemen's Land..." And in May 1848 the *West Briton* published the unhappy tale of Miss Alice Martin from Breage. Miss Martin, one of the daughters of the well-known Captain Tobias Martin of

Great Wheal Vor, had emigrated to Van Diemen's Land with her brother-in-law (a Mr. John Richards). But she had barely arrived in Hobart Town when one night she was murdered callously by a convict house-breaker. This was a salutory lesson for Cornish Methodists, who preferred the more pious (and safer!) climes of South Australia, but another reason for the greater popularity of the Central Colony in the 1840s and 1850s was that it had metal mines whilst Van Diemen's Land did not!

Indeed, Tasmania's first major mineral discoveries did not come until as late 1871, but when at last they were made they were of tin - the Cornishman's metal. Surprisingly, perhaps, Cousin Jacks in Australia had hitherto not sought extensively for tin (gold, silver, maybe even copper, were more glamorous metals), although there had been some limited tin mining in Victoria and northern New South Wales, as evidenced by placenames such as Stannum and Stannifer. Real tin mining in Australia, however, followed the discovery of the incredible Mount Bischoff deposits in the remote western highlands of Tasmania in December 1871. The find could not have come at a worst time for Cornwall, which was attempting to rebuild its sadly battered mining industry around the revival of Cornish tin, and Tasmanian competition proved almost fatal. In 1875 the *West Briton* lamented that "In the history of mining in Cornwall it has known no such disasterous year as 1874... Australia, which has been made the home of many Cornishmen, threatened at one time to annihalate altogether tin mining in this country." Two years later, in June 1877, the same newspaper expressed the view that "... a very large number of our Cornish mines must cease working..." for "... we Cornishmen have never had such a competition in the production of tin as we have at present in Australia."

But Mount Bischoff did provide employment for many emigrant Cousin Jacks. Miners who had settled in Tasmania in earlier years, with the original intention of turning their hands to other occupations, flocked to the "Mountain of Tin." And, ironically, one of the district's largest producers was itself named the "Cornwall Tin Mine." Captain William White came out from the St. Just United Mine in Cornwall to manage the West Bischoff working, and Captain W.H. Wesley - from the same parish - came out to go to another of the Bischoff mines. Others came from other parts of the continent, as in June 1885 when a party of miners arrived from Moonta and Wallaroo.

By 1878 the Mount Bischoff mines were fully developed, the incessant flow of tin ore being taken to Launceston for smelting in the six Cornish horizontal furnaces that had been erected there. Launceston was Cornish-named (though pronounced Lawn-cess-ton, clumsy when compared to the Cornish Lanson), and was situated upon Tasmania's River Tamar, in her own County of Cornwall. Perhaps the Cousin Jacks felt at home amongst the tin mines and the Cornish names, for certainly one wrote in 1914 that those "... who know anything about Latchley or Chilsworthy in eastern Cornwall..." would find many similarities between Cornish and Tasmanian scenery!

In the wake of Mount Bischoff, discoveries were made in other parts of the colony. In 1874 tin was found on the Boobyalla River in the north-east of Tasmania, while tin-seekers struck silver at Zeehan, a mining town to the south of Mount Bischoff. And in 1877 gold was found at Brandy Creek in the Tamar Valley, producing a modest rush to the area. Thus it was that Tasmania became a mining colony, and yet another Antipodean home of the Cornish miner.

Queensland, too, could boast its Cousin Jacks and metal mines. Like Tasmania, the colony yielded tin - but this time only on a small scale, following a discovery at Fishers Gully in southern Queensland, some 160 miles from Brisbane, in 1872 by four Cornishmen (originally from Redruth) who had travelled up from South Australia. It was first copper, and then (and more importantly) gold, which put Queensland on the mining map of Australia. In 1861 a rich copper deposit was found by a digger (who was searching for gold, as it happened) at Peak Downs in the central region of the colony. A handful of copper miners from Burra Burra in South Australia was hired to commence operations at the site, but it was not until the copper price rises of 1872-73 that extractive work there began on a large scale. However, in January 1868 the Adelaide *Register* had carried an article on "... the migratory tendencies of Cousin Jack...," noting that a number of miners had left Moonta for Queensland, while in the same month the *Wallaroo Times* recorded the departure of some 70 or 80 men and their families who were travelling to Peak Downs under Captain Osborne. Although a number of these miners wrote back to Yorke Peninsula complaining of fever and high prices, this did not discourage Captain Tredinnick (originally from North Downs mine, Gwennap) from leaving Wallaroo Mines for Queensland.

Nor did it prevent the Peak Downs proprieters from recruiting several hundred more miners direct from Cornwall itself. In April 1872 the *West Briton* noted the departure of 200 miners, while in the spring of 1873 another 350 emigrants left for Peak Downs - enticed by promises of wages from £15 to £20 per month. Among them was one Charles Simmons, a miner from the small village of Menheniot, near Liskeard, in East Cornwall. Simmons stayed at Peak Downs until the collapse and abandonment of the mine in 1876-77, when copper prices fell to a new low, then making his way down to Moonta Mines in South Australia. Many others went with him, among them Philip Orchard, Thomas Cock from Redruth, and John Retallick from St. Austell, with Peak Downs doomed to remain idle for ever.

It was, of course, gold that most diggers sought. And in 1867 rich deposits were located at Gympie in southern Queensland. As was so often the case, the South Australian Cornish were in the forefront of the field's development - Thomas Warn and Edward Dunstan travelling up from Kapunda as soon as the news was broadcast - but the most important movement from the Central Colony to the Gympie goldfields did not occur until the 1880s when economic pressures led the colony's Cousin Jacks to look elsewhere for work. In June and October of 1885, for example, sizeable contingents of miners were noted leaving the Moonta district for Gympie.

The discovery of the Gympie goldfields gave impetus to the exploration of Queensland's more remote areas, leading to further discoveries of gold - at Charters Towers in 1871, at Palmer River in 1873, and later at Mount Morgan in the 1880s. The Adelaide *Register* recorded the departure of Peninsula miners for Charters Towers in 1886, during that fateful decade of depression, while in 1887 the Disraeli Company advertised in the *Yorke's Peninsula Advertiser* for men to go to its workings at Rishton, in the same district. In 1907 there were still former Burra men in the area, one contributer to the *Australian Christian Commonwealth* journal in that year writing that "... in Charters Towers, Queensland, I had a hearty greeting from one of the Sampsons, of Kooringa." Certainly, the Cornish made their mark - and not least in the case of F.C.B. Vosper, from St. Dominick, who migrated to Queensland in 1883. He found his way to Charters Towers where he first became involved in politics and journalism, two activities that would bring him fame (or perhaps one should say notoriety) in Western Australia a decade later. Vosper became sub-editor of the Charters Towers *Northern Miner,* later forming his own labour weekly, the *Republican.* Through its pages he supported the Queensland sheep-shearers in their bitter strike in 1890, and he led his own miners' strike at Charters Towers, during which there was substantial rioting for which he was imprisoned for three months. Thereafter, he worked briefly on the staffs of newspapers in Sydney and Melbourne, before moving to the West in 1892.

The Palmer River Rush was the result of a deliberate attempt to seek gold deposits in the almost inpenetrable interior of northern Queensland. Men flocked to the field, but the heavy tropical rain of 1874 was to catch many of them out. A fair number were marooned by the flood waters, it being asserted at the time that several had died from starvation, so complete had been their isolation. Certainly, bands of beleagured and panicky miners ran amok in Cookstown, trying to force their way aboard the Sydney-bound steamers. One Cornish miner who had been to Palmer River sent his impressions to the *West Briton:*

Every steamer to Cook Town brings numbers of fine, strong, healthy-looking fellows, who are full of hope, and will listen to no tale of disaster, resolved upon going to the diggings to see for themselves how matters are. The same boat takes away South about the same number of men, pale, and thin, and glad to escape from the country, many of them with no money in their pockets, and few, indeed, with the gold they so eagerly seek.

It was a familiar enough tale, and had already been heard in Victoria and on the Barrier, and was later to be heard again in the West. But gold diggers never seemed to learn their lesson, as the same correspondent observed in the *West Briton:*

Still they come and still they go in hundreds from all parts of the colony, steamboat companies benefitting from the ill-wind that blows for so many. The chief thing that attracts men to the North is that it is a new country, a golden country too, and there is no knowing what riches may be found between Palmer and Cape York, a distance of about five hundred miles, where the white man has never trod - in fact, a terra incognita.

The real golden riches were found, however, in central Queensland - at a place called Mount Morgan. Attracting a number of South Australia's displaced Cousin Jacks, the mine was in production by 1886. And by 1907 the Mount Morgan company had paid a staggering £7 million in dividends. But it was not all easy going, a major metallurgical problem being posed by the difficulty of extracting the gold from the ironstone in which it was found. It was overcome, however, by the company's talented and determined metallurgist, Captain G.A. Richard, who - from his name - may well have been a Cornishman, as was certainly his assistant and assayer, Henry Trenear.

As evidenced above, the Cornish were a significant element in the mining population of Queensland. But they were by no means the majority, and amongst the other ethnic and national groups on the goldfields were a great many Chinese. Indeed, at Palmer River they completely out-numbered the whites, with 17,000 Chinese to 1,400 white men. As elsewhere in Australia, the Chinese "coolies" were feared and hated, for both cultural and economic reasons. They were aloof and alien, and their presence had a generally depressing effect on wages and conditions. But the problem did have a lighter side, as John Langdon Bonython appreciated when he wrote that "A storekeeper in Queensland with a good old Cornish name, John Tonkin, to wit, had to advertise that he hailed from Cornwall, and not from China." Others took a less humoured view of the issue, as in 1881 when one miner wrote that "I would advise... my countrymen, for I am a Cornishman, to scout these pests, and show them unmistakably that they are not welcome here." In the end the Queensland Government did just that, passing a blatantly discriminatory Act which compelled the Chinese to pay far higher mining fees than the whites.

The Northern Territory, too, had its "Chinese question." Gold had been discovered there as early as 1871, and in distant Adelaide (from which the Territory was then ruled) John Bentham Neales, from Plymouth in Devon, and several colleagues formed a company to exploit the deposit. There was fierce dealing in shares in the South Australian capital, but the mines themselves were under-capitalised and difficult to supply economically with goods and machinery on account of their remote locations. Many felt that the workings could only be made to pay if cheap labour was employed -

and that meant the Chinese. Coolies flocked into the Territory, especially after the Palmer River debacle, and by 1879 they outnumbered the Europeans by seven to one. As in Queensland, the Government acted against the "yellow peril" with discriminatory legislation, and the white miners displayed their usual contempt for the Asiatics. The Cousin Jack who wrote in 1914, after 27 years prospecting in the Territory, that "... there is no comparison between Chinese and such miners one finds at Moonta Mines ..." was in fact being quite polite! Certainly, despite the numerical dominance of the Chinese, the Cornish were in demand on account of the particular hard-rock mining experience that they possessed. Francis Manuel from St. Blazey, who had arrived in Australia in 1865, was typical of those Cornishmen who roamed the Territory, sometimes prospecting on their own account, and at others offering their services to larger concerns. At the turn of the century, when gold was found at Arltunga, in the southern part of the Territory, groups of miners were brought in from Moonta and Wallaroo.

New South Wales featured prominently in earlier chapters, first of all with the Rush of the 1850s, and then with the rise of Broken Hill. But Cornishmen were also involved in mining in other parts of the colony. In 1868, for example, a party of Moonta men went to work in the coal mines at Newcastle (although on the whole Cousin Jacks "looked down" upon coal mining, regarding it as a lesser occupation). Especially noteworthy was the rise and fall of the Great Cobar mine, which was very much a Cornish affair. Situated some 400 miles from the coast, in the interior of New South Wales, the Cobar copper deposits were first located in the late 1860s by two Danish well-sinkers. They showed their samples to a woman who had worked as a bal-maiden in Cornwall, who assured them that it was indeed copper that they had found (perhaps it is more than mere coincidence that Cobar is reminiscent of the Cornish language word for copper - cober), and they sent the ore to South Australia to be assayed.

The find was shown to be "payable," and by the early 1870s there was an influx of Cornish copper miners to the area. In December 1871 Captain Lean left Adelaide to become manager of the Great Cobar mine (or C.S.A. as it was known, short for the "Cornishman, Scotchman and Australian"), and he was joined by Captain William Morish - originally from Truro - who travelled up from Ballarat to become underground captain. Others, such as Captain Dunstan, came up from Moonta and Wallaroo, and further groups made their way to Cobar from Peak Downs. By 1882 the town of Cobar had a population of some 2,500 persons and, although some of these were lost subsequently to Broken Hill, the mine continued to prosper. Indeed, there was a boom in 1906-7 and the mine gained from the high copper prices during the First World War. But with the dramatic slump in prices that accompanied the ending of hostilities, the Great Cobar was forced to close in March 1919.

Mainland Australia, then, together with the island colony of Tasmania, was fairly covered with Cornishmen searching for metals. And the more remote elements of Australasia, while not directly within the scope of this book, are also of interest on account of the Cornish that they attracted. It is important to remember, for example, that New Zealand shared in the development of Antipodean mining and was successful in drawing Cornish emigrants to its shores. The Cornish in New Zealand is a subject that warrants a full study in its own right, but in pre-Federation days the colony was often considered the "seventh" Australian colony, and in any case there was a considerable two-way movement of Cornishmen between New Zealand and Australia.

The volume of emigration from Cornwall to New Zealand first became considerable during the 1840s, as the "Hungry Forties" began to bite. As early as 1841 a free passage scheme was in operation, and the ubiquitous Latimer and Duckham advertised through the pages of the *West Briton* and in their usual posters. "To MINERS, masons, carpenters, and all persons engaged in building, Blacksmiths, Agricultural Labourers, and Mechanics generally" they addressed their information, and it is interesting to note that even at that stage Cornish miners were in demand in New Zealand. One enterprise involved in the business of land settlement, the New Zealand Company, chartered the schooner "Regina" to convey Cornishmen to the colony during 1841. And in the following March the "Timandra" arrived with a number of Cornish on board, while the *West Briton* listed forthcoming sailings for Auckland and Wellington. In May 1843 the same newspaper noted that recent Cornish arrivals in New Zealand aboard the "Blenheim" had purchased land in the colony. By the late 1840s the pace of migration had quickened still further, lecturers at the Assembly Rooms in Truro explaining "... the capabilities and advantages of New Zealand as a field of emigration ...," the potato blight leading to a considerable exodus to the colony from West Cornwall.

California and Victoria, with their incredible gold discoveries, caught the imagination of Cornwall. But the worthy citizens of Truro were also impressed when Pake-a-Range (sic), a Maori chieftain, visited their town in January 1853 to lecture on New Zealand. And not to be outdone, New Zealand also arranged its own mineral rushes. Cornish miners had already been encouraged to migrate to the colony, and in January 1854 the *West Briton* noted that a party of Cousin Jacks under Captain Thomas Martin were at work at Kaawaa (sic, perhaps Kawhia or Kawakawa - both in North Island). Others left Cornwall to join them in the colony - Thomas Jacka from Penzance, Jonathen Uren, Phillipa Bartle from Crowan - while a further wave travelled across from mainland Australia. In July 1862, for example, Captain James Datson from Moonta visited the South Island following the gold discoveries along the Molyneaux River, spending several months prospecting at Tuapeka, Wakatipu and on the Shotover River before returning to Yorke Peninsula.

In the 1870s there was a further goldrush, to the Otago district of New

Zealand. A Cornish community grew up in the mining town of Hamiltons in Eastern Otago, and the surviving diary of one of these Cousin Jacks (Thomas Common from Stenalees, near St. Austell) forms the basis of Audrey Paterson's illuminating study *A Cornish Goldminer at Hamiltons.* Together with his colleague Silas Hore, Common emigrated to New Zealand in 1871 - visiting first the Waipori diggings, where there were a number of Cornishmen, and then travelling on to Hamiltons itself. Hamiltons was then a lively town of some 4,000 souls, a bewildering mixture of Chinese, Irishmen, and other nationalities, but in this melting-pot Common found more than a sprinkling of Cornishmen. There were miners with familiar Cornish surnames such as Thomas, Pascoe, Kinsman, and Rosevear, there were the usual Cornish wrestling matches, and the Weslyan local preachers included in their number a Cousin Jack by the name of Henry Flamank. In 1878 Thomas Common was joined by his sweet heart from Stenalees, one Elizabeth Martyn, and in the same year they were married. Later, Common went into partnership with Thomas Rosevear to open a store in the nearby township of Enfield. By then the Hamiltons mines were in decline. The major workings were abandoned by 1888, although elsewhere in the colony the goldfields survival rather longer - inevitably attracting further movements of unemployed Moonta and Wallaroo men during the difficult decade of the 1880s.

This movement from Australia to New Zealand was mirrored in a similar movement to other parts of Australasia, in particular to the French colony of New Caledonia where there were several copper mines. Captain Issac Killicoat, from Perranwell, visited the colony briefly during his time as superintendent of the smelting works at Burra Burra. And when Captain John Warren resigned his post as manager of the Hamley Mine (near Moonta) to run the Balade Copper Mines in New Caledonia he was joined by an enthusiastic band of Cornish miners anxious to journey with him - among them William Bray from St. Just-in-Penwith and William Stocker from Biscovey. Some years later, of course, Warren returned to Australia to become one of the great Broken Hill captains. Others moved on to new areas, a number finding their way to the copper mines of Papua New Guinea which as late as May 1935 still advertised in the *People's Weekly* and other Yorke Peninsula newspapers for skilled hard-rock miners.

Quite apart from the enormous geographic "spread" that the Cornish miner was able to achieve, the above - and indeed the whole Cornish "diaspora" within Australia - illustrates the remarkable mobility that the Cousin Jacks enjoyed (or perhaps one should say endured). Countless numbers spent their lives moving across the continent from one mining district to another as the fortunes of the various mines rose and fell, and as new fields were discovered. The motivation for such movement was, of course, primarily economic. But there did develop a certain quality of restlessness which spurred the Cornish on from camp to camp, from one mining town to another. The miners called this "ringing" and one old-time prospector, Fred Blakely, claimed that it was possible to observe the symptons of "ringing" as they developed: "Instead of working a man finds himself walking about the field, going from one claim to another, having a yarn with this fellow and that, between two minds what to do next." Eventually the digger would persuade one or two others to move on, and together they would leave to try their luck elsewhere. Be that as it may, this mobility was a phenomenon that was noted from the earliest days of mining in Australia. In 1846, for example, Francis Dutton showed the attraction that his Kapunda mine had had for Cornishmen, not only in South Australia, but in other parts of the continent as well. He wrote:

... Cornish miners who happen to have emigrated to the other Australian colonies, were not slow in finding their way to South Australia, to resume those occupations most congenial to the pursuits they had been accustomed to in the mother country... I may instance, in particular, two brothers of the name of Nicholls, (I believe from the parish of Gwennap in Cornwall) who obtained the first set, for the space of twelve months, at Kapunda...

This "internal mobility" was matched to a considerable extent by a Cornish "international mobility" in which Australia played a part, many Cousin Jacks spending their lives roaming, not merely within continents, but also between them! A great many Cornish, for instance, had made their way to Australia via the Americas There were a number of Californian Cornish on the Victorian goldfields in the 1850s, and Comstock silver miners went to Broken Hill in the 1880s. Inumerable individual examples could be given - Captain Henry Roach at Burra Burra had mined in Columbia, the Warmington brothers at Wallaroo and Moonta had been in the United States, and James Crabb Vercoe had worked in Mexico. There was Thomas Gregor from Cuba, George Vercoe from Wisconsin, William Moyle from Keweenaw, Richard Sloggett from Canada, Henry Crougey from Chile - all of them Cousin Jacks who had travelled to Australia by way of the New World. Men at Moonta and Kalgoorlie called them "White-washed Yanks," on account of their American accents, though John Langdon Bonython insisted that "... the twang of the American is only the Cornish accent inordinately developed."

Needless to say, there was a corresponding movement of Cornish miners away from Australia to America. Many left to become Californian "Forty-Niners," and thereafter there was a steady flow, especially to the United States. It was not altogether unusual to see newspaper notices such as that which announced "Mrs. T. Penrose and family left here (Moonta) on Wednesday morning, en route for America. They will join Mr. Penrose at Salt Lake City, their future home." The last major movement of Australian Cornishmen to North America was not until as late as 1897, when a party of

Moonta miners visited the Klondike Gold Rush. Those who left were of course enticed by stories of American affluence, but sometimes their fate was hardly enviable. Thomas Angove, for example, left Kadina for the States in 1874, but within four years he was dead. Some said that he had been gunned-down at a card gambling table, but others had heard that he had been crushed by a runaway railway truck which had careered into a gunnis (open-cut trench) in which he had been mining. Such isolated examples, however, were hardly likely to deter others from going.

As G.B. Dickason shows in his recent *Cornish Immigrants into South Africa,* southern Africa - both the Boer settlements and the British colonies, together with the Rhodesias - was yet another important home of the Cornish miner. Though few made their way from South Africa to Australia (which by then had ceased to be a principal destination for Cousin Jacks), there was in the period 1880 to 1910 a small but discernable movement of Cornishmen from Australia to South Africa. Charles Simmons from Menheniot, who had mined at Peak Downs and at Moonta, spent several years on the Rand. James Coad from Moonta worked in the Bloemfontein mines in the Orange Free State, and Nicholas Nicholls - another Moontaite - died at Boksburg in the Transvaal in 1912. By then the Cornish had acquired a truly international identity, and we find the curious example of the Moonta *People's Weekly* reprinting an article from a Cornish paper in which it was said that James and Albert Retallick - "... Australians, but of Cornish descent, their mother being a native of Redruth..." - were the champion stopers (miners) of all South Africa. In the same way, the Peninsula newspapers took a keen interest in those Cornishmen caught up in the Boer War. The proceedings of a meeting of the Ponsanooth Mutual Improvement Society, in Cornwall, were quoted at some length in the *People's Weekly,* illustrating that the Uitlanders (many of them Cornish) had received rough treatment at the hands of the Boers. And when it was alleged that the Cornish had shown great cowardice in their flight from Johannesburg, the *People's Weekly* could note that such reports were "... resented locally by Cornishmen..." and protest that the Duke of Cornwall's Light Infantry had shown great courage on the field of battle.

But however cosmopolitan the Cousin Jacks had become, nothing - in those days before Jumbo Jets and international telephone calls - could detract from what Geoffrey Blainey has called "the tyranny of distance." In migrating to Australia, the Cornish were placing themselves literally half-a-world away from home, separated by a long and often dangerous journey by sailing ship. Even those who could afford to bring their immediate family with them left behind friends, relatives, and often elderly dependents (to whom money would have to be sent), and few could have avoided the pangs of homesickness that strike even the most enthusiastic emigrant.

Correspondence with home was inevitably erratic. The "English Mail" ships were frequently delayed by weather conditions, and often the Cornish would leave only vague forwarding addresses as they moved from one mining district to another, piles of unclaimed letters accumulating in the General Post Offices in the colonial capitals. And many were illiterate, unable to read or write, relying on friends for help when dealing with correspondence - including perhaps personal and intimate details - to or from loved ones at home.

Of course, those left behind in Cornwall were no less the victims of the tyranny of distance. The *West Briton* noted in 1865 that "... the small trader is suffering greatly from bad debts, suddenly made through customers emigrating...," although also reporting that the emigrant "... Cornish miner ... rarely fails not only to benefit himself, but those of his friends remaining at home, by the welcome remittances which arrive by almost every mail." But when the flow of money dried-up, as it often did, the situation became less happy. "Colonial fever" claimed many emigrants, and not a few were the victims of mining accidents. One can only admire the stoicism and courage of Jane Berryman of Cury, on the Lizard Peninsula, who had lost not only her son but very probably her major source of income as well. Her letter, written to Immigration Officials in Australia, would have been typical of many:

Polwin, Cury, Cornwall
Nov. 18th 1857

Sir,
Yesterday I Recieved your letter informing me of the Death of my Son - I had heard the sad News some months since, and about the latter end of last September I send a letter - requesting you to send the property home - since then I have heard the expense would be considerable, for the freight. If this letter arrives before you dispose of the clothes I should be obliged to you to give Mary Hodge the silver watch and a Book title Josephus History of the Jews, I wish her to have them as a present from me.

Except, Sir my sincere thanks for your kind letter - and believe me

Yours truly

Jane Berryman

P.S. Mary Hodge is a young Female who emigrated in the same vessel and an intimate friend to the deceased.

On occasions, relatives just "vanished" in Australia without trace. A Mr. S. Morcom, who advertised in the *West Briton* in January 1864, offered to trace lost relatives in South Australia for 2s 6d, with an additional charge of 5s if advertisements were to be placed in newspapers. Even as late as 1920 such advertisements were still appearing in the Australian press, requesting information about long lost husbands. The Moonta *Peoples' Weekly* asked:

Will Gilbert Thomas, formerly a Miner, of Camborne, Cornwall, England, who left there about 30 years ago, and who is believed to have arrived in Australia about 28 years ago, or any person who can give any information regarding him, or as to his whereabouts, please communicate with the undersigned at the address mentioned below. Dated 11th August 1920.
A.J. McLachlan, Solicitor, Adelaide.

Shortly before her death in the early 1920s, an elderly lady in Helston dictated a short autobiography to one George Jose (a former Moonta miner), part of which dealt with the effects of her husband's disappearance in Australia:

Many times I have gone to the wash tray without breakfast, and my two dear children have had to stay until I came home without any food in the house. I have cried myself to sleep many a night. One night I lost heart, so I took my two children to a water shaft at Bassett mines, with the intention of drowning them and myself; but the captain of the mine saw me just in time. I told him my trouble and he wrote to Captain Hancock in Australia, and found out my husband was working there. He could not do anything, but asked him to write. Captain Hancock's daughter sent me £1. I often wished I could write; I would thank her so much.

But not all were so unlucky. Many were able to keep in touch, and to continue to receive their all-important monetary assistance, and others were eventually reunited with their relatives. Typically, the Cornish miner in Australia would save as hard as he could to bring his whole family out to the colonies, so that they could all begin life anew. But there was a substantial minority who decided to return to Cornwall as soon as they had made what they considered a goodly sum. When James Penn Boucaut returned on holiday to Cornwall in 1892 he came across an old man in Saltash who had been in South Australia in the 1840s. And John Langdon Bonython, on a visit to Cornwall, met a man in St. Keverne who had spent some 25 years in Victoria before deciding to return home. At Pensilva there was a Mr. Williams - "an old Moonta identity" - and at St. Ann's Chapel there lived a Mr. Preston, also a former Peninsula resident. And at Calstock there was Captain Dunstan, an ex-Moonta miner who had lived the life of a recluse after his return from Australia. At the turn of the century he was "discovered" by Thomas Cowling, an Australian of Cornish descent, who had deliberately sought him out. Cowling wrote,

I spoke of Yelta, Paramatta, Wandilta and half a dozen other mines as though they were all within the parish of Calstock. Following a few more remarks on my part he became talkative and said: "Evidently young man, you have spent some time in Australia, for you have mentioned names of places not unknown to me."

Through his reserve, we see in Dunstan's guarded comment an affection for Australia, or at least a certain nostalgia for the time he had spent there. Others were less inhibited - in 1921 George Jose of Illogan wrote to the *People's Weekly* from Cornwall, declaring "Oh! how I fancy the old place and my school days at Moonta Mines... I left Moonta in 1902, and I should like to be back in those good old days again..." A few were even more attached to their memories of Australia, a kind of homesickness in reverse which plucked them from their retirement in Cornwall and returned them once again to the land "Down Under." Such was the case of Edward Constantine who had emigrated to South Australia from Cornwall in 1874, working in the Wallaroo smelting works until his retirement in 1899, whereafter he returned to Cornwall. But he missed his friends, his children who had settled out there, the long hot summers... and so he went back to the country that had become his real home, living quietly in the Adelaide Hills until his death in 1934. Slowly, subtly, the Cornish were becoming Australians - assimilation was at hand.

...... AND ENDINGS

By the turn of the century Cornish miners had found their way into most corners of the Antipodes, or at least to those districts where there were metals to be had, but already the assimilatory process was well in hand. Where they were only small minorities, the Cornish were soon absorbed into the broad mass of British-Australians, and even where they were represented in some number - as at Bendigo, Broken Hill, and Kalgoorlie - their identity was blurred by an emerging Australian consciousness, though retaining perhaps a "tincture of Cornishry" as A.L. Rowse has called it. Certainly, in the first two decades of this century, the Cornish in Australia became much less an identifiable ethnic group, as their cultural identity was progressively eroded.

But there was one area where it was not all so simple, where the process of assimilation was subtler, more lengthy and more complex - it was, predictably, the northern Yorke Peninsula of South Australia, the last remaining focus of Cornish culture in the continent. Even there significant cultural change was noticable by the turn of the century. Memories of Cornwall became ever more romantic and nostalgic, many years having passed since the reality of migration from hunger and unemployment, while in the public mind actual qualities of "Cornishness" were associated with life on the Peninsula itself, rather than with distant Cornwall. Moonta and environs, instead of Cornwall, became the focal point of Cousin Jack loyalty and identity. And a rather Janus-like culture emerged on the Peninsula - one which continued to look back to Cornwall, but was dynamic and vibrant enough to develop along its own independent course.

Countless examples of the survivals of Cornish culture could be quoted, especially from the local newspapers, but the rate and nature of cultural change is more difficult to pin down. To a considerable extent, though, in a rather symbolic way, this change was personified in H. Lipson Hancock, the son of the celebrated Captain Hancock, his own attitudes and behaviour mirroring the cultural changes that were occurring. He was born and bred on the Peninsula, spoke with a pronounced Cornish accent, and was generally considered a "Cousin Jack." One of his colleagues wrote that,

Lipson Hancock had all the Cornish virtues and only one of the vices. He was warm-hearted, hospitable, and genuinely concerned for the welfare of those under him... Lipson's one limitation - and he knew it - was that of suspiciousness, characteristic of Cornishmen.

But despite this apparent "Cornishness," H. Lipson Hancock regarded himself first and foremost as an Australian. This did not prevent him from making the traditional pilgrimage to Cornwall, but when he arrived there his impressions were exactly congruent with that Janus-like culture from which he sprang. On the one hand, he experienced feelings of affinity and "belonging," while on the other he knew that Cornwall - a land he had never seen before - could never really be "home" for him. His sketch of Launceston on market-day says it all:

... the crowds of country folk in the streets called up memories of home people at Moonta on Saturday nights. In fact, there is a strong feeling of being as near home in that district as it was ever possible to be, away from the true "homeland" by the sea in sunny South Australia.

Like many Cousin Jacks, H. Lipson Hancock was a mining man and an official in the Methodist Church, and, in accordance with the Cornish tradition, took a keen interest in mining welfare policies. However, in each of these areas he adopted attitudes and programmes which were to some extent novel, and diverged from normal Cornish practice.

In the mining sphere, for example, he was a "book learned" qualified engineer, in contrast to his father and the earlier generation of "practical miners" who had worked their way up to be captains. And while many Cornish managers tended to be conservative, but with a flair for improvisation (the cynics would say "bodging"), Lipson Hancock was a meticulous and methodical perfectionist - a man almost obsessed with a concern for order, rationality and progress. He had assumed full-time management of the Moonta and Wallaroo Mines in 1895, and when his father retired completely in 1898 Lipson was confirmed in the post of General Superintendent. In only a few years he transformed the workings from typical, rather old-fashioned Cornish mines to modern industrial plants - the old Cornish engines and their stone-built engine-houses, and the flimsy "shears" standing above the shafts, giving way to enormous corrugated-iron power-houses and gigantic head-frames. There was a considerable movement of men and materials from Moonta to Wallaroo (by now the more important of the two mines), and much of the investment undertaken after 1900 was concentrated on the latter mine. Like his father before him, H. Lipson Hancock was often prepared to flout and ignore the wishes of his Directors in his relentless bid to modernize and expand. Again reminiscent of his father, he eventually bullied the Directors into submission, the Board becoming little more than a "rubber stamp" for his bold and ambitious policies. And with the aid of two boom periods, 1905-7 and 1911-18, company profits since amalgamation in 1890 reached £1,301,000 by 1918.

Associated with his mining policy, but again reflecting the contradictory influences in his up-bringing, was Lipson Hancock's "Betterment Principle" - the formalising and rationalising of the company's hitherto ad hoc welfare policies into a unified, coherent set of regulations and practices. He felt that the coercion implicit in both Factory Legislation and trade union activity was unjustified, and certainly an affront to liberal values, and argued instead that the onus was on the employer to provide satisfactory working conditions and amenities for his workers. He felt, however, that the Moonta and Wallaroo welfare system was untidy and in need of review, and by 1912 he had formulated his "Betterment Principle" - a comprehensive scheme which he and others outlined in a series of articles and pamphlets. The system provided employees with everything from recreational facilities to subsidised building lots, and included a sophisticated Club and Medical Fund. In all, it was the most advanced welfare organisation to be found anywhere in Australia at that time.

And if the "Betterment Principle" was a model of clarity and order, then H. Lipson Hancock's "Rainbow System" of Sunday School instruction was little short of authoritarian rigidity. Hancock was Superintendent of the Moonta Mines Sunday School, and in 1905 he was given permission to implement his American-derived "Rainbow System." His books on the subject are full of key words such as "re-organization" and "reconstruction," illustrating again his love of progress and providing further insights into his personality. The School, when re-organised, consisted of ten Grades, ranging from the "Cradle Roll" from children from infancy to three years, to the "Home Grade" for those too old or infirm to attend. Although the emphasis was naturally on the children, there was a "Senior Grade" for those from 16 to 23, and an "Adult Grade" for people aged 23 and upwards. To administer the system there were no less than 22 different committees, while the actual "Rainbow" course of instruction consisted of a number of pre-determined lessons, commencing with "Lesson 1: The Bible and how we got it" and culminating in "Lesson 100: The Call of China." The actual practice of the Sunday School was a rigid and systematic as the theory, the meetings being conducted according to a regimented and unvarying code of behaviour. One visitor to the School remarked, with a mixture of admiration and amazement, that at each meeting,

A large card suspended before the superintendent's desk bore the legend 'I am early.' A minute or so prior to the opening of the service this was replaced by another bearing the words 'I am only just in time.' When the service began, a third card appeared announcing 'I am late'... Among the mottoes on the walls, 'Study to be quiet' was prominent.

Although in many ways novel, the "Rainbow System," being somewhat stern and puritanical, was not out of character with the broad stream of Victorian-Edwardian non-conformity. However, it was never really in tune with traditional Cornish Methodism, which tended to be more exuberant and emotional - as in 1905 when there was a great Revival in the Moonta circuit, one observer writing:

Glory! Glory!! Glory!!! Glory be to the Father, and to the Son,
and to the Holy Ghost! Yes, we have had the Barrett Brothers
on the Mines and have had a glorious time. If you don't believe
it, come and see. Oh, such a revival!

Whoever the evangelical "Barrett Brothers" were is a question lost to history, but the kind of fervid Methodism that they preached had a lasting effect. South Australian Methodism, particularly its Cornish strain, was then still linked with liberalism and democratic socialism, and the Cornish Radical Tradition found its greatest expression in the State Premiership of "Honest John" Verran in 1910 to 1912.

Verran was born in Gwennap, Cornwall, in July 1856 and as a youngster was employed as an ore pickey-boy (sorter) at Kapunda. As a youth he moved to Moonta Mines, where he worked underground until his election to the South Australian State Parliament in 1901. He was a Methodist local preacher, and to those who argued that religion had no place in politics, he would reply:

Religion is citizenship, and the relationship between religion
and politics is very close... When we come to justice and
righteousness and truth... religion is not just a question of going
to heaven. It is a question of living and making the world better
for having been in it...

In April 1910 the United Labor Party, with Verran as its leader, was swept into office for the first time, pledged to a programme of wide reform. The Peninsula went wild with rejoicing, but the euphoria was illusory for Verran's term of office was characterised by the series of crises in which he found himself.

1910 was a year of industrial strife in South Australia, and Verran's opponents were able to make great capital out of his apparent sympathy for the trade unionists. In the December the Adelaide cart-drivers struck, an event which led to street violence. Verran narrowly escaped dismissal by the State Governor, but the damage to his party and Government had already been done. And the opposition were given further ammunition when, in 1910-11, Verran acquired several of the minor Yorke Peninsula copper mines to work as State enterprises. Quite apart from the supposed regional "favouritism," this "nationalisation" was seen as evidence of Verran's committment to advanced socialism, although it was a genuine desire to promote the mining industry rather than ideological dogma which lay behind his actions. Verran's handling of the crisis was arrogant and inept, and he further enraged opposition opinion by purchasing two more mines without reference to Parliament.

In addition, the opposition was able to exploit Verran's practice of "tacking" - the attaching of provisions for new, hitherto undiscussed, public works to the Appropriation (Budget) Bill - which they saw as an attempt to push through socialist legislation without proper Parliamentry scrutiny. The conservative upper house (the Legislative Council) had already rejected a Bill to limit its own power, so it needed no persuading to also reject the Appropriation Bill of December 1911 (which included "tacked" provisions for the establishment of a State brickworks). Verran made a brief and unsuccessful appeal to Asquith for Imperial intervention but, with his Supply refused, he had little choice but to call an election. The Government was by now discredited badly, and was of course heavily defeated - the only constituency not to follow the anti-Labor trend being the mining district of Wallaroo. Verran was personally discredited by the defeat of February 1912, and he resigned from the Labor leadership in the following year.

Another consequence of Labor's defeat was the estrangement of the Methodist Church from the United Labor Party. The Methodist *Australian Christian Commonwealth* magazine had earlier supported Verran, but claimed in February 1912 that he and his colleagues had been too "... prepared to take their orders from the more violent and revolutionary forces in their party" and that "... efforts are being made to dominate the Labour Party by the Church of Rome..." This estrangement was made all the more complete by the events of the Conscription Issue and the Great War, and was traumatic in the extreme for the Cornish community - faced as it now was by divided loyalties.

The Conscription Issue was an Australia-wide phenomenon, and although the anti-conscription stance was essentially a conscientious one, it was seen by many as an Irish Catholic and pro-German plot. Unlike its eastern States counterparts, the United Labor Party in South Australia had few Irish (or even Catholic) members, but there were many Prussian and Silesian settlers in the State, and when the ULP declared against conscription the worst fears of its opponents seemed to be confirmed. The Methodist Church, with its strong anti-Catholic prejudice, declared in favour of conscription, and so its estrangement from Labor became absolute. Verran, in the event, resigned from the ULP to form the Moonta Branch of the *National* Labor Party (pro-conscriptionist) and he was able to retain the support of many local people. But others, the anti-conscriptions, found a new champion in Robert Stanley Richards, a young trade union activist.

Richards was born at Moonta Mines in 1885, the son of Richard Richards of Camborne and Mary Jeffery from Tuckingmill. He was, paradoxically, a Methodist local preacher but politically he was noticably to the left of Verran and the earlier, Cornish-born generation of miners. Standing together with John Pedler, another local of Cornish descent,

Richards defeated Verran and his "running mate" Herbert in the Wallaroo Constituency in the 1918 elections. Verran, out of Parliament for the first time since 1901, needed desperately to win new sources of political support, and he did this by embracing the German question. He argued that German settlers in the State should be disenfranchised. "I am a Britisher and a Cornishman," he said, "It is deplorable to allow those with German blood in their veins to vote in this country." For a time this campaign gained him some support, but after the War the issue lost its immediacy and Verran failed to regain his seat in 1921. Thereafter, he continued to drift to the right, and in 1924 contested Wallaroo as a Liberal, declaring that "He could not see how any country could accept socialistic proposals..." But yet again he was beaten by Richards and, with the exception of a brief spell in the Federal Senate from 1927-28 when he was appointed to fill a casual vacancy, his political career was finished. He died in 1932.

The "split" in the Labour Party in South Australia over the Conscription issue was mirrored in a corresponding "split" in the miners' union on Yorke Peninsula. R.S. Richards and the younger militants argued that their branch of the Amalgamated Miners' Association should merge with the Australian Workers Union, for "unity is strength," but Verran and the older Cornishmen insisted that such a move would lose local control over industrial matters. The merger went ahead, but Verran retaliated by forming a new union, known to its opponents as the "Bogies." The Wallaroo and Moonta company was accused of victimising the militants who had engineered the AWU merger, while the "Bogies" were in turn victimised by the AWU men and ostracised by the local community. And despite years of relative industrial calm, this "split" of 1917 ushered in a new era of conflict and confrontation. This culminated in a brief but bitter strike in 1922, led by R.S. Richards, when the AWU's venom was directed equally at the company and the "Bogies."

When the mines closed in 1923 there were those who blamed the collapse on the AWU, while there were certainly some militants who were pleased to see the company go at last into liquidation. By now, of course, the old Cornish tradition was sadly battered - the linkage of Methodism with social democratic liberalism was broken, the sense of solidarity had gone, and a new extremism had emerged. The dispersion of the local population after the closure helped further to erode the Cornish influence in local politics, although during the late 1920s and early 1930s there remained in the Adelaide Parliament a small core of Labor men of Cornish birth or descent - Thomas Gluyas, Stan Verran, Thomas Hawke, Albert Hawke, Thomas Edwards, Leslie Hunkin, Henry Kneebone, Stanley Whitford. R.S. Richards himself became State Premier, though for only 64 days prior to the election in 1933 when the Labour Party was swept from office. As before, Labor had lost its credibility - but this time as a result of internal disagreements as to whether the Party should adopt the controversial "Premiers' Plan" (devised by New

South Wales) for economic recovery. Thereafter, little of the Cornish influence remained in South Australian politics, and the Wallaroo constituency became only a shadow of its former self as more and more people left the district.

The Conscription Issue and its aftermath, then, had led to the disintegration of the Cornish Radical Tradition in the State. Other facets of South Australia's Cornish inheritance were also altered and diminished by the effects of the First World War, an experience which perpetuated, accelerated, and accentuated the pace of cultural change. It is often said, with some justification, that the spirit of Australian nationhood was born on the heights of Gallipoli. Young soldiers from Yorke Peninsula shared in this new sense of identity, and were as quick as any others to criticise the "Pommy" troops and Generals. But still they had not entirely exchanged old "roots" for new, as evidenced in war-time letters sent home from Europe.

Private Leigh Lennell, for example, was a young ANZAC trooper from Moonta Mines. He and his pal Art Trenwith (another Moontaite) were wounded at the Front and sent to a nursing home at Putney, where Lennell had an artificial arm fitted. As soon as they were fit enough, they went on a convalescing tour of Cornwall. In early 1917 Lennell wrote from the "Tywarnhaile Hotel," Perranporth, to his parents at Moonta Mines:

You will see by this address that I am down with the Cousin Jacks. It is perfectly lovely here, and just like home again. The people talk exactly like the Cornish at Moonta. The scenery and sights are simply beautiful, and as soon as the people knew we were Australians and of Cornish descent, they crowded around us and talked for hours ... I feel sure some of them must have relatives in Moonta. Such names as Polkinghorne, Pengilly, Polgreen, Penberthy, are all folk here. Today we went to Bedruthan Caves and to St. Agnes tin mines. I feel quite at home here.

As he had anticipated, Lennell did "encounter" people with relatives in Moonta - messrs. Rowe and Penberthy, who ran an outfitters' in Penzance and still received copies of the Moonta *People's Weekly!*

That Lennell's case was not an unusual one, nor his enthusiasm atypical, is demonstrated by the very similar letters sent home to Moonta by Private Lloyd Pollard when he was on leave in Cornwall in 1916. He wrote:

At last I can write to you about my trip through dear old Cornwall - the places you used to speak of when Roy and I were boys. Then we used to hear dad and grandfather speak of the places I had the pleasure to see. The places they used to mention came back as clearly as if I were quite familiar with them...

208

Pollard wrote with obvious pride about holding "... my first conversation with a Cornishman in his native country..." and said that "It was grand to hear the Cornish dialect..." Soon after he returned to France, and, alas, two years later he was dead - fatally injured while stretcher-bearing under fire.

But while there were those to whom it was important to make these sentimental journeys to Cornwall, there were other soldiers from the Peninsula who were content to merely insist to all and sundry, be they British or Australian, that "If you haven't been to Moonta you haven't travelled" and that Moonta was "The Hub of the Universe." These men, perhaps, were exhibiting the other face of the Peninsula heritage - the intense local patriotism in which Moonta had replaced Cornwall as the object of Cousin Jack loyalty and identity.

After the War the assimilatory process continued apace, the great catalyst being the closure of the Moonta and Wallaroo Mines and the consequent dispersion of the Peninsula population. With the end of the War in 1918, the demand for copper had fallen dramatically, causing copper prices to slump. There were several periods of inactivity at the mines after March 1919, before they were abandoned finally on 23rd October 1923, the Wallaroo and Moonta Mining and Smelting Company going into liquidation in the November. Many had anticipated the crash, realising that the mines could not survive the combination of low prices, soaring costs and growing impoverishment of the lodes, and as early as 1919 many had left the district - bands of unemployed miners gathering in Adelaide, where they created comment by singing their hymns and carols. Others tried to find work in areas such as Broken Hill and Kalgoorlie, or in the smelting works at Port Pirie.

It was mainly the young and the fit who left the district (thus robbing it of much of its life and vitality), leaving behind a high proportion of elderly and infirm people. But despite the general air of desolation and gloom, many of the "old timers" clung with surprising tenacity to their cottages on the mineral leases. And, inevitably, there were those who insisted that the mines had been closed prematurely, and that hitherto untouched ore-bodies were still awaiting exploitation. In 1924 one "... reputable miner of 55 years' experience in Moonta Mines, in addition to the Burra, Kapunda, Wallaroo Mines, Broken Hill, and Cornwall..." argued that the mines could still be made to pay, given appropriate government support.

And while a handfull of the more enterprising locals leased sections of the old mines on a form of tribute, with a Moonta Copper Recovery Company reworking the waste tips, the State Government did indeed move in to try to promote a revitalisation of the industry. In 1929 the Minister of Mines, none other than R.S. Richards, arranged a system of State subsidies for a Moonta Prospecting Syndicate, while in the following year he also supported the more ambitious Moonta Mines Development group. By June 1932 some 62 men were employed in the various Moonta ventures, with head-frames again dominating the local sky-line, but by 1938 each of the schemes had failed, with

209

the workings abandoned for good.

Northern Yorke Peninsula was no longer a copper mining district, all that was history. And with the death of the mines came the end of a living Cornish culture, and its replacement by a rather self-conscious sense of identity, one that could be "turned-up" to suit particular occasions. One such occasion was the "Back to Moonta" celebration of 1927, when many of those who had been forced to leave the district came back to remember the old days. The editor of the Moonta *People's Weekly* wallowed in the nostalgia of it all, declaring:

The Celtic spirit is deep set in folk that hail from Cornwall, and they are Celts on an equality with the Scots, Irish or Welsh. They are clannish to a marked degree, and the love of home and the clan seldom loses its hold on the individual and never in the race, and Moonta people are mostly Cornish.

The *People's Weekly,* indeed, became the principal vehicle of this increasingly sentimental view, not only of Cornwall, but also of the Peninsula's past, publishing such articles as "St. Piran: The Patron Saint of Cornish Miners" and "Cornish Nationality," confident that "Our community has been in every sense a bit of old Cornwall... Moonta, in its proximity to the sea, is able to carry the role of an outpost of the delectable Duchy."

But, however eloquent, the relevance of such prose to the everyday life of the Peninsula people was probably marginal. Fortunately, other local writers were busy recording what life for the Cornish settlers in the district had really been like. Phyllis Somerville produced a prize-winning novel *Not Only In Stone,* an authentic account of early Peninsula life, and Oswald Pryor poured out an incredible volume of Cousin Jack stories and cartoons, culminating in the publication of his delightful local history *Australia's Little Cornwall* in 1962. The son of Captain James Pryor from Wendron, Oswald Pryor (born at Moonta Mines in 1881) was, more than anyone, responsible for preserving and popularising - through his prolific work - the essence of Cornish life on Yorke Peninsula.

Other enthusiasts worked through the medium of the Cornish Association of South Australia, the Penzance *Cornishman* noting in 1919 that:

... the remarkable feature is that the love and enthusiasm for the county extends unabated to the third and fourth generations of Cornish Australians. The Association's membership contains a long list of distinguished South Australians... of persons who hail from Cornwall or have Cornish blood in their veins.

Sir John Langdon Bonython, who died in 1939 at the great age of 91, was first and foremost amongst these distinguished enthusiasts. He had amassed a great personal fortune as owner of the Adelaide *Advertiser* newspaper, and had briefly represented the State as a radical-Liberal politician in the Federal Senate. He succeeded Lord Falmouth as President of the Royal Institution of Cornwall in 1931, declaring that he believed King Arthur really would return and that, despite the Anglo-Saxon conquest of England, "... Cornwall continued to remain a country apart." For his services to Cornwall, Sir John was created a Bard of the Cornish Gorseth.

The Cornish Association survived through the Depression years of the 1930s, when it was beset with financial problems - a situation made worse by the fact, as the Association's minutes noted, that ... the younger generation had no interest in the old Land or its traditions, having been born here with different surroundings." But there were those who cherished their Cornish connections and descent, and after the Second World War they were joined by the sprinkling of Cornish amongst the new wave of British migrants to the State, revitalising the Association to the point where - in 1977 - it was able to inaugurate a series of bi-annual Cornish seminars. Inspired by this rejuvenation, the near-defunct Sydney Cornish Association was reborn with new vigour, and a similar "West Country" group in Victoria was able to attract new Cornish enthusiasts to its ranks.

This rekindling of interest was consistent with the Post-War Australian search for "roots" and historical identity, and found its greatest expression in the "Kernewek Lowender" Cornish festival - a lavish event which has been held bi-annually on northern Yorke Peninsula since 1973. The accent is on authenticity, and the visitor may wander through the ruined engine-houses and spoil tips, munching his pasty and swigging his "swankey," watching the black and white Cornish flags fluttering from public buildings in the distance, and hearing the strains of the haunting Furry Dance tune - played by a local brass band - as they drift across the old mineral leases. It is a reminder of how deep is the Cornish heritage in Australia, and recalls the old adage "What do they know of Cornwall, who only Cornwall know?".

FINIS

APPENDIX A

FAMILIES, Cornish Emigrant:

Adams 37,81,107; Alford 178; Allen 115; Andrewartha 36; Angove 101,167,198; Angwin 167; Annear 30; Anthony 49,50; Archer 52; Arthur 177; Axford 34; Badcock 119; Bargwanna 75; Barker 52; Barkla 14,38,49; Barnett 183; Bartle 195; Bastian 19,34,41,126,167,178; Bath 35,112; Batten 13,40; Beaglehole 106; Begelhole 183; Begolhall 114; Benalack 32; Bennett 36,120,145,170; Bennetta 84,85; Bennetts 13,85,144; Berriman 167; Berryman 144,186; Besanco 36; Bice 42,107; Bishop 167; Blamey 82; Blatchford 80; Blewett 112; Bligh 9; Blight 41; Bluet 180; Boaden 20; Boase 119; Bone 41,113; Bonython 69,70,200; Bosance 95; Boswarva 36,95; Boucaut 28,69,70,92,93,98,99,104,105,107,200; Bowden 44; Bray 21,41,43,51,52,76,79,101,111,139,167; Britton 144; Broad 9; Brokenshire 144,81; Bryant 9,35,39,50,113,114,132; Bunney 42; Burnard 147; Burrows 165,167; Carceek 112; Carlyon 119,121; Carthew 33,113,142,143; Champion 131; Chapell 143; Chapman 41; Chappell 44; Charleston 106,107,170; Chellew 120,121; Chenoweth 127; Chenery 126; Christopher 34,114; Clymo 112; Coad 131,198; Coca 82,191; Cocking 95,185; Coles 52; Colwell 71; Common 196; Congdon 35,37; Constantine 201; Cooke 116; Coon 13; Cornelius 43,114; Cornish 39,144; Cowling 43,44,200,201; Crewes 147,148; Crougey 197; Curgenwin 169; Curnow 22,38,41,112,120,128; Curtis 50; Daddow 170; Dalley 118; Daniel 120; Datson 49,133,195; Davy 37; Davey 18,32,144,167,182; Deason 126; Deeble 87; Doble 50; Doney 76; Downs 172; Drew 148; Dunn 15; Dunstan 37,45,47,113,114,144,145,148,166,183,191; Dunstone 41,115,167; East 44,45,47,96,182,184; Eddy 107,120,145,172; Ede 33; Edwards 42,144,207; Ellery 41; Ellis 51,119,145,177; Faull 87; Flamank 196; Fletcher 14; Foster 86; Francis 183; Gilbert 167; Giles 120-123,128; Gill 86,96,98,99,101,105; Glasson 14,41,111,119; Gluyas 207; Goldsworthy 24,35,42,167; Goyne 119; Gray 42,147; Gregor 197; Grenfell 13,111,131; Grigg 22; Grose 106; Gummow 34,113, Gundry 22; Halse 80,82; Hambly 183; Hamblin 141; Hammer 119; Hancock 21,51,101,128,139; Harris 34,36,43,114; Harvey 120,121,143,178; Hawke 37,40,111,112,131,145,167,188,207; Hayes 75,87; Hicks 52,145,167; Higgs 42,46-48,126; Hoare 144; Hocking 143,145,187; Hodge 69,199; Holman 15; Hooper 22,106,107,143,148; Hore 196; Horton 182; Hosken 84,86,94,95; Hosking 117,120; Hunkin 207; Hutchins 33; Inch 112; Isaac 143; Isaacs 43,168; Ivey 171; Jacka 195; James 87,122,129, 167,182; Jeffery 124,126,206; Jeffree 112; Jenkin 32,52,82,113,131,143,166; Jenkins 112; Jewell 44,126; Johns 167,175; Jones 37; Jose 200; Jury 80; Karkeek 32; Kellaway 143,167; Kelynack 119; Kemp 126,143; Kendall 165; Kent 24; Kerby 144; Kestel 34,113; Killicoat 40,196; King 9; Kinsman 30,196; Kitto 42,96; Kneebone 187,207; Knowles 96; Knuckey 33; Lander 96, 167; Lane 111; Langdon 131; Lanyon 131; Lathlean 180; Laurimer 13,52; Lawry 120; Lean 22,33,42,50,194; Lennell 208; Lethlean 20; Lewis 119; Liddicoat 20,44,182; Lobb 112; Luke 139; Madden 44; Maddern 107; Magor 167; Major 127; Manuel 43,194; Martin 19,37,120,127,144,178,189,195; Martyn 196; Mathews 118,119; Matthews 52,120,124,128, 129,143,144; May 37,146,147; Mayne 45; McArthur 93,99; Medland 14,18; Medlyn 73; Menhennett 51; Mildren 115; Mill 20; Mitchell 33,35,76,95,126,144,148,174,186; Morcom 34,80; Morish 144,169,194; Morrison 50; Morton 182; Moyle 20,22,40,43,95,124,131,197; Moyses 41,167; Murrin 101; Mutton 44; Nancarrow 167; Nankervis 40; Nankivell 167,182; Newton 175; Nicholas 81; Nicholls 31-34,39,50,94,112,119,197,198; Ninnes 41,78,172; Noble 43; Northey 42,112,126,182; Oates 15,18,44,111,126,145,165,176,186; Odgers 33,115,167; Olds 172; Oliver 40,127,143; Orchard 15,20,52,191; Osborne 35,96,167,191; Parkin 49; Parkyn 181; Pasco 112; Pascoe 33,38,42,50,81,119,167,182,183,196; Paull 14,32,38,49,50,51,111; Pawley 126; Paynter 36,39,49,80,166; Pearce 40,81,92,144,175,176,183,187; Pearse 111; Pedler 13,115,206; Pellew 80,148; Penberthy 38,50,145; Pengelly 71,101,148; Penglase 145; Penglaze 35; Penhall 38,112,148; Penna 95,167; Penrose 87,197; Pergam 143; Peters 115,144,168; Philips 114; Phillips 36,37,42,49,113,139,182; Pinch 36,41,52; Piper 30,142,144,146,165,169; Polkinhorn 127; Polkinghorne 95,126,144,170; Polkingthorne 145; Pollard 126,167,186,208,209; Polmear 120; Poole 51,139,140; Pope 86; Preston200; Prideaux 42,167,172; Prior 169; Prisk 30,49,97,98,100,105,167; Prout 33,124,126,138; Prowse 19; Pryor 43,47,77,97,167,177; Pyatt 142; Quick 117; Quintrall 36; Read 120; Rendell 19; Repper 20; Retallick 43,144,147,191,198; Richard 193; Richards 19,20,24,35,75,101,105,108, 144,190,206,207,209; Roach 35,40,47,120,126,197; Roberts 14,34,35,51,112,114,139,143, 148,182,183; Robins 19; Rodda 17,38,41,49,80,119,131,182,183; Rogers 50,95,114,144,167;

213

Roscrow 36; Rose 44; Rosevear 196; Rosewall 178; Rosewarne 141,144; Rounsevell 105; Rowe 21,37,38,40,71,80,82,88,101,105,124,126,131,143-145,168,180,182,185; Rowett 114, 143; Rowland 20; Rowse 167; Rule 40,112; Rundle 17,167,180; Sampson 41,130,144,192; Samson 20,101; Samuels 182; Sanders 28,40,83; Sando 44; Sandow 113,167; Santo 35,52,81, 92,107,116; Sara 148,167; Saunders 175,185; Sawle 16,52; Scaddan 187; Scoble 16; Scown 15,143,167; Semmens 122; Sibley 144; Simmons 100,121,191,198; Skewes 50; Skewis 117,124; Slee 172; Sleep 13,15,127,141,144; Sloggett 197; Smith 182; Snell 42,50,117,143; Sowden 86; Spargo 16,17; Sparnon 84; Spry 37; Stephens 9,12,92,95,101,124; Stevens 145; Stocker 196; Tambling 114; Tamblyn 43; Teague 41,114; Teddy 81; Terrell 50; Thomas 33,36,37,111-113,120,126,144,145,171; Tom 110,111; Tonkin 38,49,112,145,193; Toy 119; Tozer 143; Trebilcock 100,167,182; Tredinnick 167,191; Tregea 17; Tregloan 131; Treglown 112; Tregoning 43; Tregonning 166; Tregoweth 51,145; Tregurtha 117; Trelease 44;Treleaven 20; Treloar 20,113,122,123,138,167,178,183,184; Tremaine 131,144; Trembath 40,144,172; Tremelling 167; Tremewan 36; Trenberth 167; Trenear 193; Trenery 40; Trenerry 167; Trenomon 147; Trenouth 42; Trenowden 32; Trenowith 17; Trenwith 208; Treseder 122; Tresise 86,139,140; Tresize 20,145,167; Trestrail 39,97,112,143,185; Trethowan 33; Trevaskis 40; Trevean 95; Trevellyan 125; Trevena 34,112,114; Trevorrow 36; Trewartha 33,95; Trewavas 118; Trewenack 107; Trewick 144; Trewidden 29; Trewin 143,166; Trezona 44; Triplett 143; Truran 42; Truscott 34,182; Tyrell 34; Uren 112,167,195; Veale 33; Verco 52,92,116,197; Vercoe 42,43,49,197; Verran 108,182,205,206,207; Vial 39; Visick 98,105; Vivian 80,113; Walter 42; Warmington 47,96,97,197; Warn 191; Warren 48,144,146,148,183,196; Warwick 182; Waterhouse 92; Waters 34,40,114,137; Wearne 42,121; Wesley 190; Whitburn 143,174; White 120,126,145,190; Whitford 114,182,187,207; Wills 69; Williams 13,101,120,139,144, 145,178,179,185; Withington 86; Yelland 36; Yeoman 20,87.

Mines, Cornish:
Basset 200; Burra Burra 16; Caradon 25,36; Consolidated 21,25; Crinnis 38; Devon Great Consols 25,147,184; Dolcoath 25,30,145-147; Drakewalls 44; East Wheal Rose 16; Fowey Consols 27; Great Wheal Busy 27,171; Great Wheal Vor 113,145,190; Levant 175,185; North Downs 191; North Pool 171; Old Moor 49; Old Treburgett 50; Pendeen Consols 171; Perran Great St. George 34; Polberrow 36; Retanna Hill 43; St. Day United 27; St. Ives Consols 172; St. Just United 145,190; Tresvean 25,35,40; Trewavas Head 113; United 36; Wheal Burton 14; Wheal Edward 44; Wheal Elizabeth 49; Wheal Lovell 43; Wheal Owles 176; Wheal Rock 34; West Kapunda 16.

Australian:
Adelaide Sovereign 183; Albion 42; Alpha 112; Anaconda 145; Arrow Proprietary 183; Associated 177,182; Australian 182; Badra 174; Bayley's No. 1 South; Bailey's Reward 183; Beltana 51; Benalack's 32; Bendigo Coolgardie 179; Big Hill 145; Block 10 144,146,165; Block 11 172; Block 14 182; Bon Accord 37; Bonanza 145; Botallick 146; Botallock 32; Boulder Central 183; Britannia 146; British 144; Broken Hill Central (Central Mine) 144,147; Broken Hill North (North Mine) 141,144; Broken Hill Proprietary 141,144,146,148,172; Broken Hill South (South Mine) 141,144,146,172,184; Broughton River 49; Brown Hill 183,184; Burra Burra 16-18,22,27,28,33-41,49,50,71,72,80,83,92,94,98,99,103,113,116,123, 127,128,131-133,135,136,209; Callington 32,37,38,43,72; Carn Brea (S.A.) 32; Carn Brea (N.S.W.) 146; Carrick Range 113; Central Wealth Consolidated Goldfields 181; Charlton 49; Copper Valley 42; Colonial Goldfields 183; Consols 177; Cornish 111; Cornish United 131; Cornwall 131; Cornwall Tin 190; Crinnis 32,38; Dalcooth 146; Davey's 32; Day Dream 140; Devon Consols 184; Ding Dong 32; Doora 42; Duke of Cornwall 32; Eagle Hawk 145; Easter Gift 183; Ediacara 50,182; Emu Flats 37; Enterprise 33; Euko 42; Faith 182; Fraser South 176; Geraldine Lead 174; Gipsy Girl 145; Glen Osmond Union 33; Great Boulder 177,182,188; Great Britain 145; Great Cobar (Cornishman, Scotsman & Australian) 144,145,174,194; Great Coolgardie 183; Great Devon Consols 32; Great Northern 50; Great Norther Junction 141; Great Wheal Grey 35; Great Wheal Orford 37; Greenock Creek 38; Hamley 42,44,48, 144,196; Hannan's Central 182; Hannan's Consols 183; Hidden Secret 145; Hit and Miss 183; Irish Lily 183; Iron Duke 177,182; Ivanhoe 182,188; Jenkin's 32; Junction 144,145; Kalgoorlie Reef 183; Kanmantoo 37,38,43; Kanowna Consolidated 183; Kapunda 25,27,34,36, 38,39,41,83,94,131,132,135,136,205,209; Karkarilla 42; Karkulto 37; Kirkeek's Treasure 32; Kurilla 9; Lady Berys 145; Lady Looh 183; Lake View 177,182,187; Lazarus New Chum 129; Leigh Creek 51; Lyndock Valley 38; MacFarlane's 33; Malcom Proprietary 182; Maritana 182; Mc. Auliff's Reward 184; Mochatoona 50; Montacute 34,35,80; Moonta 182; McAuliff's Reward 184; Mochatoona 50; Montacute 34,35,80; Moonta 27,30,31,33,42, 44 - 47,49,77,87,96 - 98,102,103,107,127,133,136,142,146,147,168,172,177,188,194,198,203 - 205,209; Mount Lyndhurst 51; Mount Remarkable 50; Mount Rose 51; Mount Tincroft 146; Mukurta 34; Murninnie 49; Murray Scrub 37; Mutooroo 51,143; Napier 182; New Cornwall 32,42,48,76,96; New Mile 145; North Grenville 127; North Kapunda 22,34; North Montacute 34,110; North Rhine 38; North Yelta 44; Northern Wealth of Nations 181,182; Old Cornwall 32; Old Geraldine 174; Paramatta 42,76,201; Pararra 51; Paringa 114; Paull's Consolidated 32; Peak Downs 191, 198; Phoenix 32; Pine Creek 175; Pinnacles 140,144; Poona 42,43; Port Lincoln 49; Prince Alfred 50; Princess Royal 35,37; Purnamoota 144,168; Reedy Creek 37,38 72,107; Reefer's Eureka 184; Richmond Gem 183; Rise and Shine 145; Rising Sun 145; Rockwell Paddock 145; Round Hill 144; Royal Mint 177,182; Sliding Rock 50; Smith's Olary 51; South Kapunda 34; Strathalbyn 114; Sons of Gwalia 188; Talisker 145; Terrible Dick 145; Three Australias 177; Tincroft 146; Trengoff 32; Trenowden's Claim 32; Tresevean 32; Trevue 32; Trident 145; True Blue 183; Truro 32; Umberumburka 138,140,142,146; Umberumburka Extended 139; United Hills 32; Victoria Cross 141,144; Victoria Deep Quartz 129; Victory 145; Vocovocanna 50; Wallaroo 27,29 - 31,42,44 -48,74,96 - 98,102,103,132,133,136, 142,144,147,172,177,182,188,203,204,209; Wandilta 42,201; War Dance 145; West Bischoff 145,190; West Umberumburka 139; Wheal Acraman 37; Wheal Barton 38,50; Wheal Bessie 49; Wheal Blinman (Blinman Mine) 50,73,98; Wheal Boone 37; Wheal Burrawing 49; Wheal Burty 42; Wheal Byjerkno 146; Wheal Charles 34; Wheal Devon 42; Wheal Dutton 34; Wheal Ellen (S.A.) 38,39,46,80; Wheal Ellen (W.A.) 174,184; Wheal Fanny 38; Wheal Fortune 38; Wheal Friendship 32,38; Wheal Gawler 16,32; Wheal Gundry 34; Wheal Hardy 33;

Wheal Hughes 42,44; Wheal James 42; Wheal Margaret 49; Wheal Maria 38; Wheal Prosper 22,32; Wheal Rose 38; Wheal Rothschild 37; Wheal Virgin 32; Wheal Watkins 33,38; Wirrawilka 50; Wirtaweena 51; Yelta 42,44,182,201; Zinc Corporation 147.
Other:
Comstock 147,197; Balade Copper 196; Real de Monte 118.

APPENDIX C

SHIPS:
Andromanche 111; Aldinga 133; Ascendant 23; Amoor 20; Augusta Schneider 118; Blenheim 195; Bounty 9; Britannia 17; Canterbury 20; China 17; Collingrove 20; Coorong 133; Dalhousie 28; Eliza 18; David Malcom 17; Eagle 111; Gosforth 27; Great Britain 120; Henry Wellesley 89; Himalaya 18; Hooghly 17,24; Isabella Watson 16,21; Java 19,20; John Brightman 52; Kangeroo 133; Kingston 17; Lady Ann 20,24; Lady Milton 27; Lillies 21; Marion 20; Mystery 119; Nile 23; Omega 23; Orleans 189; Pakenham 18; Prince Regent 18,21; Princess Royal 17,21; Queen Bee 27; Rajah 16; Red Admiral 13; Regina 195; Snowdrop 119; Stag 22; Sultana 23; Tarquin 27; Theresa 17; Timandra 111,195; West Australian 28; Westminster 20; William Metcalfe 111; William Money 18; William Prowse 23.

Select Bibliography

ARCHIVAL SOURCES:

a) *Charles Rasp Memorial Library (Broken Hill)*
Amalgamated Society of Engineers, Broken Hill Branch, Minute Book 1891-96.
Barrier Miner Business Directory 1891
Broken Hill South Church - Wesleyan Methodist Roll Book 1892-1912.
Wesleyan Marriage Register for Silverton and Broken Hill 1886-1893.

b) *Cornwall County Record Office (Truro)*
DD BRA 1678/8/9 & 10*Letters from Richard Hancock to his Mother,* 1862-64.
DD BRA 1678/11   *Letter from John Pascoe to William Hancock,* September 1864.
DD CN 3426   *Letter from John Glasson to Col. Angus Carlyon,* August 1850.
DD CN 3427/1   *Letter from William Pomeroy Carlyon to his parents* February 1853.
DDX 384/3   *Kilkhampton Bible Christian Circuit Minute Books* 1891-1900.

c) *Helston Museum*
2305 *Copy of a letter from Joseph Orchard,* July 1848.
2290 John Boaden, *Some Account of My Life and Times,* 1902-4.

d) *La Trobe Library (Melbourne)*
Despatch: Cornwall and Devon Society to Young, *Memorial as to Emigration of Cornish Miners,* Despatch No. 19, 31 January 1851.

c) *Mitchell Library (Sydney)*
A38091/2 *Carlyon Papers* (originals in *Cornwall County Record Office*)

f) *Royal Institution of Cornwall (Truro)*
Extracts from the *West Briton* and *Royal Cornwall Gazette* relating to Cornish emigration in the nineteenth-century.
Posters relating to emigration from Cornwall to Australia.

g) *South Australian Archives (Adelaide)*
94   *Letters, Chiefly Commercial, to James & Robert Frew*
97/379-98u   *Boucaut Papers*
124   *Report on the Bible Christians of Australia* c. 1881.
313   *Passenger Lists*
522   J.G. Bice, *Draft of First Hustings Speech,* March 1894.
581s   *Notes by L.C.E. Gee concerning welfare work at the Wallaroo and Moonta Mines*
842m   *Notes on the Burra Mine: Diary of Johnson Frederick Hayward* 1846-56.
1209   Rodney Cockburn, *Nomenclature of South Australia,* (revised version).

217

1509/1510	*Burra Mines: Sundry Documents 1849-61*
1529	*Alphabetical Index to Applications for Free Passage from the United Kingdom to South Australia 1836-40.*
3105L	*Letter from Richard Geo. Clode to his mother, September 1852*
4959L	E. Major, *Notes on Moonta and Wallaroo*
A3 98/A3	*Testimonial provided for J.G. Bice by Capt. H.R. Hancock, June 1876.*
A856/B9	*Historical and Descriptive Account of the Burra Mine, 1881.*
A858/B4	A.A. Lendon, *The Formation of the South Australian Mining Association.*
A1118	*A Holograph Memoir of Captain Charles Harvey Bagot.*
BRG 22	*South Australian Mining Association Papers.*
BRG 40	*Moonta and Wallaroo Mines Papers.*
D2631L	*An Anonymous Diary 1850-1851.*
D31961T	Oswald Pryor, *The Glen Osmond Mines and the Presence of Cornish Miners there 1841-51*, 1953.
D3217T	Oswald Pryor, *Little Cornwall: The Story of Moonta (South Australia) and its Cornish Miners.*
D3458T	K.T. Borrow, *The Glen Osmond Mines*, 1955.
D3627L	Stanley R. Whitford, *An Autobiography.*
D4718L	George Richards, *Journal of a Voyage to South Australia.*
D48001	*Abstract from Diary of the late Francis Treloar.*
D4876 Misc	W. Shelley, *The Great Strike*, c1874.
D51049L	*Letter from John Whitford, Kangaroo Flat, Victoria to his father and family*, 1852.
D5133 (Misc)	*Musical Scores of hymn tunes by James Richards.*
D5569T	J. Harbison, *Historical Notes on Moonta Cemetery.*
D5865T	W.H. Hayes, *Notes on Moonta Mines.*
D6010 Misc	Max Slee, *Accidental Mining Deaths at Moonta and Wallaroo Mines 1866-1900*, 1977.
D6018T	Meryl A. Kuchel, *Pies and Pasties: A Cornish Community in an Australian Environment*, 1976.
D6029/1-115L	*Letters written Home by Cornish folk who emigrated to Australia in the nineteenth-century* (collected by J.M. Tregenza, includes material from *Cornwall County Record Office*).
GRG 24	*Correspondence to the Colonial Secretary.*
GRG 35/43	*Department of Lands Records, Immigration Agent*, 1849-78.
PRG3	*Leaworthy Papers.*
PRG 81	*Robert Stanley Richards Papers.*
PRG 86	Thomas Burgess, *Papers relating to Immigration c1843-49.*
PRG 96	*Oswald Pryor Papers.*
PRG 174	*George Fife Angas Papers*
PRG 341	*Letters written by Daniel and David Williams to their parents while travelling overland from Wallaroo to Perth through the goldfields 1894-96.*
RN 260	*Some Notable Cornish Pioneers.*
SRG 4/34/1	*Broken Hill and Silverton Wesleyan Circuit Baptismal Registers 1890-93.*

SRG 4/35/1   *Kooringa Chapel Quarterly Minutes and Leaders Meetings 1848-55.*
SRG 4/1-3/1   *Diary of the Rev. John G. Wright 1856-1901.*

h) *University of Melbourne Archives*
Broken Hill Mine: Misc. Minutes and Correspondence.
Creswick Primitive Methodist Church Quarterly Meetings 1885-1902.

NEWSPAPERS AND PERIODICALS:

a) *British*
Bible Christian Magazine
Commercial, Shipping and General Advertiser for West Cornwall
Cornishman
Cornish Magazine
Cornish Telegraph
Devonport Telegraph
Mining Journal
Old Cornwall
Primitive Methodist Magazine
Royal Cornwall Gazette
South Australian News
South Australian Record
Wesleyan Methodist Magazine
West Briton

b) *Australian*
Australian Christian Commonwealth
Barrier Miner
Barrier Truth
Bathurst Free Press
Burra News
Burra Record
Christian Weekly and Methodist Journal
Kalgoorlie Miner
Kalgoorlie Western Argus
Kapunda Herald
Northern Mail
Northern Star
Observer (Adelaide)
People's Weekly
Plain Dealer
South Australian Advertiser
South Australian Dept. of Mines Mining Review
South Australian Gazette and Colonial Register
South Australian Primitive Methodist
South Australian Register
Wallaroo Times
Yorke's Peninsula Advertiser

BOOKS, PAMPHLETS & ARTICLES:

"A Circle of Friends"*Memories of Kapunda,* Kapunda Herald, Kapunda, 1929.

Harry Alvey,   *Burra: It's Mines and Methodism,* South Australian Methodist Historical Society, Adelaide, 1961.

Anon,   *The Barrier Silver and Tin Fields in 1888,* 1888, republished, Libraries Board of S.A., Adelaide, 1970.

Ian Auhl,   *Burra Sketchbook,* Rigby, Adelaide, 1969.

     and Dennis Marfleet,   *Australia's Earliest Mining Era: South Australia 1841-1851,* Rigby, Adelaide, 1975.

J.B. Austin,   *The Mines of South Australia,* 1863, republished, Libraries Board of S.A., Adelaide, 1968.

D.B. Barton,   *Essays in Cornish Mining History, Vol. 1,* D. Bradford Barton, Truro, 1968.

     The Cornish Beam Engine, D. Bradford Barton, Truro, 1968.

     Life in Cornwall in the Nineteenth Century, (4 Vols.), D.

Bradford   Bradford Barton, Truro, 1970-74.

John Blacket,   *History of South Australia,* Hussey & Gillingham, Adelaide, 1911.

Geoffrey Blainey,   *The Rise of Broken Hill,* Macmillan, London, 1968.

     The Rush That Never Ended; A History of Australian Mining, Melbourne University Press, Melbourne, 1963.

     The Tyranny of Distance, Sun Books, Melbourne, 1966.

Fred Blakely,   *Hard Liberty,* Harrap, London, 1938

W.L. Blamires and John B. Smith,   *The Early Story of the Wesleyan Methodist Church In Victoria,* 1886.

John Langdon Bonython,   *Cornwall: Interesting History and Romantic Stories,* Advertiser, Adelaide, 1932.

J.P. Boucaut,   *Letters to My Boys.* Gay & Bird, London, 1906.

F.W. Bourne,   *The Bible Christians,* Bible Christian Book Room, London, 1905.

C.W. Bowden,   *History of Agery,* Bowden, Agery, 1966.

H.Y.L. Brown,   *Mining Records of South Australia,* 1887

     Records of the Mines of South Australia, S.A. Govt. Adelaide, 1908.

H.T. Burgess,   *The Cyclopedea of South Australia,* (2 Vols.), Alfred G. Selway, Adelaide, 1907.

Roger Burt (Ed.),   *Cornish Mines and Miners,* D. Bradford Barton, Truro 1972.

     (Ed.),   *Cornish Mining,* David & Charles, Newton Abbot, 1969.

Thomas Burtt,   *Moonta Musings in Rythmic Rhyme,* 1885.

Arnold Caldicott,   *The Verco Story,* Caldicott, Adelaide, 1970.

Gavin Casey and Ted Mayman,   *The Mile That Midas Touched,* Rigby, Adelaide, 1964.

D.M. Charleston,   *Address to the Electors of Port Adelaide,* 1906.

     New Unionism, 1890.

Robert Charlton,   *The History of Kapunda,* Hawthorne, Melbourne. 1974.

Donald Clark,   *Australian Mining and Metallurgy,* Critchley Parker, London, 1904.

E.H. Coombe,   *History of Gawler 1837-1908,* Gawler institute, Gawler, 1910.

L.S. Curtis,   *The History of Broken Hill,* 1908, republished, Libraries Board of S.A., Adelaide, 1968.

Frank Cusak,   *Bendigo: A History,* Heinemann, Melbourne, 1973.

Michael Davitt,   *Life and Progress in Australasia 1898.*

Graham B. Dickason *Cornish Immigrants to South Africa,* Balkema, Cape Town, 1978

Francis Dutton,   *South Australia and Its Mines,* 1846.

Una R. Emanuel,   "The History and Decline of the Mining Village of Byng" *Australian Geographer,* Vol. 1., Part 2, November 1929.

Jean Fielding and "The English of Australia's Little Cornwall,"
W.S. Ransom.   *Journal of the Australasian Universities Language and Literature Association, No. 36,* November 1971.

James Flett,   *The History of Gold Discovery in Victoria, Hawthorne,* Melbourne, 1970.

Stanley G. Forth, *Methodism in the Clare District,* South Australian Methodist Historical Society, Adelaide, 1974.

L.E. Fredman,   "Sir John Quick's Birthplace," *Victorian Historical Journal,* Vol. 48., No. 1., February 1977.

Thomas Gill,   *History and Topography of Glen Osmond, 1905,* republished Libraries Board of S.A., Adelaide, 1974.

W.R. Glasson,   "Early Western Glimpses," *Journal and Proceedings, The Australian Methodist Historical Society,* No. 43., January 1945.

H. Lipson Hancock, *Modern Methods in Sunday School Work,* Methodist Book Depot, Adelaide, 1916.
                   The Rainbow Course of Bible Study, 1919.
                   "Welfare Work in the Mining Industry", *Australian Chemical Engineering and Mining Review,* October 1918.

Milton Hand,   *Moonta, Wallaroo, Kadina Sketchbook,* Rigby, Adelaide, 1974.

Charles G. Harper, *From Paddington to Penzance,* 1893.

Peter Heydon,   *Quiet Decision: A Study of George Foster Pearce,* Melbourne University Press, Melbourne, 1965.

Samuel Higgs,   "Some Remarks on The Mining District of Yorke's Peninsula, South Australia," *Transactions of the Royal Geological Society of Cornwall, Vol. 9, No. 1.,* 1875.

A.D. Hunt,   *Methodism Militant: Attitudes to the Great War,* South Australian Methodist Historical Society, Adelaide, 1976.

W.F. James,   *Joseph Hancock, Nonegerarian Methodist Minister,* Adelaide, 1914.

J.B. Jaquet,   "Early Days at Barrier Ranges," *Geology of the Broken Hill Lode and Barrier Ranges Mineral Field,* 1894.

Fred Jones   *The Hon, Sir Langdon Bonython,* Royal Cornwall Polytechnic Society, Camborne, 1931.

R.H.B. Kearns,   *Broken Hill 1883-1893,* Broken Hill Historical Society, Broken Hill, 1973.

Nancy Keesing (Ed.),*History of the Australian Gold Rushes,* Hawthorne, Melbourne, 1971.

W.S. Kelly,	*Early History of the Kapunda Methodist Circuit,* South Australian Methodist Historical Society, Adelaide, 1959.
Brian Kennedy,	*Silver, Sin and Sixpenny Ale: A Social History of Broken Hill 1883-1921,* Melbourne University Press, Melbourne, 1978.
W.B. Kimberly,	*History of Western Australia,* 1897.
J.R. Leifchild,	*Cornwall - It's Mines and Miners,* 1857, republished, Frank Graham, Newcastle, 1968.
George E. Loyau,	*Notable South Australians,* 1885, republished Austaprint, Adelaide, 1978.
	The Representative Men of South Australia, 1883, republished Austaprint, Adelaide, 1978.
D.W. Meinig,	*On The Margins of Good Earth,* Seal Books, Adelaide, 1970.
John Meredith & Hugh Anderson,	*Folk Songs of Australia,* Ure Smith, Sydney, 1973.
W.P. Morrell,	*The Gold Rushes,* Black, London 1948, 2 cnd Ed, 1968.
W. Frederick Morrison,	*The Aldine History of South Australia,* 1890.
Jean V. Moyle,	*The Wakefield: Its Water and Wealth,* Moyle, Riverton, 1975.
Cyril Noall,	*Tales of the Cornish Fishermen,* Tor Mark Press, Truro, 1970.
R. Norris,	"Economic Influences on the 1898 South Australian Federation Referendum," in A.W. Martin (Ed.) *Essays in Australian Federation,* Melbourne University Press, Melbourne, 1969.
J.J. Pascoe,	*History of Adelaide and Vicinity,* 1901, republished, Osterstock, Adelaide, 1972.
Audrey Paterson	*A Cornish Goldminer at Hamiltons.* Otago Heritage Books, Dunedin, 1980.
Philip Payton,	*Pictorial History of Australia's Little Cornwall,* Rigby, Adelaide, 1978.
Douglas Pike,	*Paradise of Dissent: South Australia 1829-57,* Longmans, London, 1957.
J.C.C. Probert,	*The Sociology of Cornish Methodism,* Cornish Methodist Historical Society. Truro, 1971.
Oswald Pryor,	*Australia's Little Cornwall,* Rigby, Adelaide, 1962.
	Cornish Pasty: A Selection of Cartoons, Seal Books, Adelaide, 1976.
John Reynolds,	*Men and Mines: A History of Australian Mining 1788-1971,* Sun Books, Melbourne, 1974.
James Richards et al,	*The Christmas Welcome: A Choice Collection of Cornish Carols,* 1893.
Emilie Robinson,	*Cap'n 'Ancock: Ruler of Australia's Little Cornwall,* Rigby, Adelaide, 1978.
Gordon Rowe,	*Methodism on the Copper Mines of Yorke Peninsula, South Australia,* South Australian Methodist Historical Society, Adelaide, 1951.
John Rowe,	*Cornish Methodists and Emigrants,* Cornish Methodist Historical Society, Truro? 1967.
	The Hard-Rock Men: Cornish Immigrants and the North

	American Mining Frontier, University of Liverpool Press, Liverpool, 1973.
A.L. Rowse,	*The Cornish in America,* Macmillan, London, 1969.
Thomas Shaw,	*A History of Cornish Methodism,* D. Bradford Barton, Truro, 1967.
John Sherer,	*The Gold-Finder of Australia,* 1853, republished, Penguin, Ringwood, 1973.
Phyllis Somerville,	*Not Only in Stone,* 1947, republished, Seal Books, Adelaide, 1973.
	"The Influence of Cornwall on South Australian Methodism," *Journal of the South Australian Methodist Historical Society,* Vol. 4., October 1972.
W.G. Spence,	*Australia's Awakening,* The Workers' Trustees, Sydney, 1909.
John Stephens,	*The Land of Promise,* 1839.
Ralph S.G. Stokes,	*Mines and Minerals of the British Empire,* Arnold, London, 1908.
"Cousin Sylvia,"	*Homing,* Hassell Press, Adelaide, nd.
H.R. Taylor,	*The History of the Churches of Christ in South Australia,* Evangelical Union, Adelaide. 1959.
W.R. Thomas,	*In the Early Days: A faithful account of the Barrier Silver Field,* 1889.
K.W. Thomson,	"The changes in function of former mining settlements: the Wallaroo copper belt," *Proceedings of Royal Geographical Society of Australia, South Australian Branch,* 56: 47-58, 1955.
	Universal Depression: It's Causes and Cures, 1895.
A.C. Todd,	*The Cornish Miner in America,* D. Bradford Barton, Truro, 1967.
	The Search For Silver: Cornish Miners in Mexico 1824-1947, Lodenek Press, Padstow, 1977.
F.E. Treloar,	*The Burra Mine: Reminiscences of Its Rise and Fall 1845-1877,* Burra Record, Burra, 1929.
Seymour Tremenheere,	"Notice respecting the Lead and Copper ores of Glen Osmond Mines, three miles from Adelaide, South Australia," *Transactions of the Royal Geological Society of Cornwall,* Vol. 6., 1841-46.
Anthony Trollope,	*Australia,* 1873, republished, University of Queensland Press, St. Lucia, 1967.
P.W. Verco,	*Masons, Millers and Medicine: James Crabb Verco and His Sons,* Verco, Adelaide, 1976.
May Vivienne,	*Sunny South Australia,* Hussey & Gillingham, Adelaide, 1908.
Elizabeth Warbuton,	*Old Stradbroke,* Lynton, Adelaide, 1976.
Royce Wells,	"Early Mining in the Adelaide Hills," *Journal of the Historical Society of South Australia,* No. 2, 1976.
Michael Williams,	*The Making of the South Australian Landscape,* Academic Press, London, 1974.
O.H. Woodward,	*A Review of the Broken Hill Lead-Silver-Zinc Industry. West* Publishing Corp., Sydney, 2cnd Ed., 1965.

223

R.M. Yelland,   *Colonists, Copper and Corn in the Colony of South Australia 1850-51,* Hawthorne, Melbourne, 1970.

UNIVERSITY OF ADELAIDE UNPUBLISHED B.A. HONOURS AND OTHER THESES:

John Barrett,	"The Union of the Methodist Churches in South Australia," 1955.
I.J. Bettison,	"Kapunda: A Study of The Establishment of a Community in Rural South Australia, 1960.
K.R. Bowes,	"The Moonta Mine 1861-75", 1954.
K.W.A. Bray,	"Government-Sponsored Immigration into South Australia 1872-86", 1961 (M.A. thesis).
Henry Brown,	"The Copper Industry of South Australia: An Economic Study," 1937, revised 1960.
Mel Davies,	"The South Australian Mining Association and the Marketing of Copper and Copper Ores 1845-77", 1977 (M.A. thesis).
P.L. Edgar,	"Sir James Penn Boucaut: His Political Life 1861-75", 1961.
S.G. Fitzgerald,	"Half A World Away: South Australian Migration 1851-1872," 1969.
J.W. Higgins,	"The South Australian Mining Association 1846-77", 1956.
W.C.R. Jaques,	"The Impact of the Gold Rushes on South Australia 1852-54", 1963.
R.J. Miller,	"The Fall of the Verran Government 1911-1912." 1965.
Philip Payton,	"The Cornish in South Australia: Their Influence and Experience From Immigration to Assimilation 1836-1936," 1978 (PhD thesis).
B.M.H. Reynolds,	"Immigration into South Australia 1829-52", 1928.
Janet Scarfe,	"The Labour Wedge: The First Six Labour Members of the South Australian Legislative Council," 1968.

MISCELLANEOUS SOURCES:

Autobiography of Thomas Cowling Jnr, (private unpublished MS).
Burra Burra Mine - Copper Ore Day Book 1860-61, (Burra National Trust Museum).
Adrian Lee, *How Cornwall Votes,* (unpublished paper, Plymouth Polytechnic, 1977).
Minutes of the Cornish Association of South Australia.
Newsletters of the Cornish Association of South Australia.
Minutes of the South Australian Conference of the Methodist Church of Australia.
Miscellaneous Annual Reports of the Missionary Society Under the Direction of the Bible Christian Conference.
South Australian Almanacks, Directories and Gazeteers.
South Australian Parliamentary Debates
South Australian Parliamentary Papers.

Notes

1. *Victoria County History, Part V Romano-British Cornwall*. 1924.
2. A.G. Langdon, *Victoria County History, Early Christian Monuments*. 1906. Opinions expressed in the text about a few of these inscribed stones are the author's and are purely speculative. The period in which the stones are believed to have been erected seems to be isolated and unrelated to the periods immediately preceding and following it. It is probably the least documented, archaeologically, of the recognised periods in Cornwall's history.
3. Rev. F.C. Hingeston-Randolph, editor, *Register of Bishop Grandisson*, 1894. This letter of Grandisson is significant since it explains why the lives of the Welsh, Irish and Breton saints who came to Cornwall have been relegated to the realms of legend and fancy. It is incredible that no church in Cornwall or Devon has preserved in its chest the record of the life of its local patron saint, assuming of course that the bishop's directive was heeded.
4. R. Doehaerd, *L'Expansion Economique Belge au Moyen Age*. 1946. There is, so far, no authentic evidence of shipments of tin from Cornwall or Devon to the Continent, direct or via London, in the post-Roman pre-Norman period. However, the virtually uninterrupted tradition of Belgian metal-working means that supplies were obtained. These could only have come from the already existing source.
5. German Bapst, *L'Etain*. 1884. This French historian gives in his book what is probably the most complete account ever written of the vast range of products and articles made of tin or incorporating it, from the earliest uses down to its applications in the late Middle Ages. Although Bapst is not concerned with the commerce in tin his work implies that it must have been extensive and enduring. Only a selection of the metal's manifold uses are mentioned in this essay.
6. Germain Bapst, *L'Etain*. 1884.
7. *The Great Rolls of the Pipe, 7-9 Ric. I and I John*.
8. *The Calendar of Close Rolls, 6 Edward II*.
9. *The Calendar of Patent Rolls, 21 Edward III*.
10. A. Schaube, *Handelsgeschichte der Romanischen Voelker des Mittelmeergebietes bis zum Ende der Kreuzzeuge*. 1906. Schaube is only one of many continental historians who have noticed and reported the existence of the international trade in Cornwall's tin in the Middle Ages. In his book he gives us glimpses of its Mediterranean ramifications, whilst the Italian Francesco Pegolotti, a prominent member of the Bardi Society in the 14th century, takes

us further afield to the starting points of the penetration of the trade by Italian mercantile companies into distant Asia.

11. Schaube A. *Handelsgeschichte . . . etc.*
12. Francesco Pegolotti, *La Pratica della Mercatura*, editor Allen Evans, Medieval Academy of America, 1936.
13. Francesco Pegolotti, *La Pratica della Mercatura . . . etc.*
14. H.B. Walters, *The Church Bells of England.* 1912.
15. *The Calendar of Liberate Rolls. 31—36 Henry III.* Restriction of space prevents a more detailed and broader selection of purchases of tin by the Crown for specific applications. Acquisitions of the metal for use in such government works as building and bell-founding continued all through the Middle Ages.
16. *The Calendar of Liberate Rolls, 32—46, Henry III.*
17. Charles Welch, *History of the Worshipful Company of Pewterers.* 1912.
18. Germain Bapst, *L'Etain.* 1884.
19. Germain Bapst, *L'Etain.* 1884.
20. Jules Balasque, *Etudes Historiques sur la Ville de Bayonne.* 1862.
21. Jules Balasque, *Etudes Historiques . . . etc.*
22. *Ministers' Accounts, Duchy of Cornwall, 14th Century.*
23. *Register of Edward the Black Prince, folio 124.* HMSO 1931.
24. P. Studer, editor *The Port Books of Southampton 1427—30.* 1913.
25. L.M. Midgley, *Ministers' Accounts, Earldom of Cornwall 1296—7.* 1945.
26. L.M. Midgley, *Ministers' Accounts . . . etc.*
27. Maurice Heresford, *New Towns of the Middle Ages.* 1967.
28. T.F. Tout, *Chapters in the Administrative History of Medieval England, vol. 5.* 1930.
29. G. Concanen, *Ministers' Accounts, Duchy of Cornwall, 1349*, cited in the Trial at Bar Rowe v. Brenton. 1830.
30. Rev. F.C. Hingeston-Randolph, editor *Register of Bishop Grandisson.* 1894.
31. *Ministers' Accounts, Duchy of Cornwall, Rolls No. 6 & 7, SC 6 816—823.* These annual Ministers' accounts reveal that the adverse effects of the Black Death were felt until the end of the 14th century and beyond. They also show that the subsequent plague of 1361 caused the complete cessation of commercial fishing at five traditional fishery porths on the coast of West Penwith.
32. *Register of Edward the Black Prince, folio 100.* HMSO 1931.
33. Matthias Dunn, *The Migrations and Habits of the Pilchard*, a lecture delivered at the Fisheries Exhibition, Truro in 1895.
34. *Victoria County History, The Fisheries*, 1906.
35. *Parliamentary Papers, vol. XI, 1857.* Report of House of Commons

select Committee on the Rating of Mines.

36. This essay is the author's severely condensed account of James Silk Buckingham's Autobiography that was published in two volumes and covered his life up to the age of twenty-nine years. He died as soon as this first part of his account of his travels and adventures was published. He was unable to commit to paper the story of the rest of his life.

37. Boase & Courtenay, *Biblioteca Cornubiensis, vols. I & III.* 1874 & 1882.

AUTHOR'S POSTSCRIPT

1. Sources

Full documentation appears in the author's PhD thesis, but here most sources are indicated within the text itself, while the select bibliography lists those principal primary and secondary sources employed specifically in the writing of this book. The bibliography, therefore, should not be regarded as exhaustive; and there will, of course, be other important sources in both Cornwall and Australia which the author was not able to unearth during the period of his researches.

2. Surnames

Surnames are important historical eveidence, and anyone attempting to study the Cornish overseas should have a good working knowledge of Cornish surnames. "Exclusively Cornish" names are mainly Celtic and, as well as including the obvious Tre, Pol and Pen examples, embrace more obscure names such as Uren, Medlyn and Retallick. "Typically Cornish" names are for the most part the common patronymics - Williams, Roberts, Thomas, Richards, and so on, which perhaps the majority of Cornish people bear-though some patronymics (for example, Jenkin, Bennetta, Santo) are so closely identified with Cornwall that they must in all conscience be termed "Exclusively Cornish". Research is complicated, of course, by the fact that a number of Cornish people bear English-style occupational or descriptive names (e.g. Smith, Carter, White, Brown) while there are more than a few long-established but nevertheless "non-Cornish" surnames in Cornwall, such as Buckingham, Bishop and Warmington. By far the best guide to Cornish surnames is G. Pawley White's **A Handbook of Cornish Surnames**, 1972, (recently republished by Dyllansow Truran), while Piers Dixon's **Cornish Names**, 1975, and the surprisingly useful Cornish section of H.B. Guppy's **Homes of Family Names in Great Britain**, 1980, are also of great value.

3. Recent Research

Since the research for this book was completed, several new works have appeared on the scene. Jim Faull's **Cornish Heritage: A Miner's Story**, Adelaide,1979, is a well-researched biography-cum-family history of Captain Christopher Faull, from Crowan, one of Yorke Peninsula's prominent mining captains in the last century. Jim Faull's **The Cornish in Australia**, A.E. Press, Melbourne,1983, is essentially a summary of existing works - particularly his **Cornish Heritage** and the present author's PhD thesis - though it does offer some new material. Keith Skues' recent volume, also entitled **Cornish Heritage**, is a deeply-researched Skues family history - of interest in this context as several folk of the name Skues, Skewis,Skewes emigrated to Australia in the last century.

4. Further Research

Keith Skues' work is also of value in that it sets the Cornish "international mobility" in its correct context. To date, historians have studied Cornish emigration with regard to particular countries or continents - a broader view of the ebb and flow of Cornish migration to and between those various countries is now required. But, that said, there is still plenty of scope for further research within Australasia. Queensland and Tasmania deserve greater treatment than this author was able to give them, and we still await a full-scale study of the Cornish in New Zealand. The author's forthcoming **Cornish Carols from Australia** (due to be published shortly by Dyllansow Truran) looks at one aspect of the Cornish cultural inheritance in South Australia, and his projected study of Cornish colonists and the expansion of Adelaide and the South Australian agricultural frontier will investigate the impact and activities of those settlers who were not part of the copper mining communities.

Aberdeen (S.A.) 71
Abraham Roberts & Co. 129
Adams, Henry 37,107
Adams, Henry (Jnr) 107
Adelaide 12,16-20,22,23,28,29,33-35,37,
   44,47,50,52,69,70,74,79,81,82,85,92,93
   99,102,104,106,110,113-116,127,128
   132,134-136,138,139,141,143,145-147,
   167,170,175-176,182,184,187,188
   193,200,205,207,209,211
Adelaide Hills 33,34,37,39,134,176,201
Adelaide Mining Co. 34
Adelaide Prospecting Syndicate 176
Adelaide School of Mines 49
Adelaide United Trades and Labour
   Council 102
Albany 176,178,184
Alford, Richard 178
Allen, John 115
Allen, William 18,116
Allum, Frank E. 184,185
Altarnun 13
Amalgamated Miners' Association 102-
   104,106,130,168,169,186,207
Amalgamated Mining Managers'
   Association of Australasia 168
Andrewartha, Thomas 36
Angas, George Fife 17,38
Angove, J. 101
Angove, Thomas 198
Annear, William 30
Anthony, Cpt. 49,50
Apoinga 37,38,50
Archer, George 52
Ardrossan (S.A.) 51
Arltunga 194
Asquith 206
Aukland 195
Austin, J.B. 49
Australian Labor Party 170
Australian Communist Party 170
Australian Workers' Union 207
Ayres, Henry 22-24,27,36-40,95,132
Axford, Thomas 34

Badcock, Richard 119
Badcock, William 119
Bagot, Charles 33,34
Baldhu 44
Ballarat 43,76,112,117,119,120,124,126-
   129,132,133,144,175,194
Barbican, The (Plymouth) 17
Barker, Thomas 52
Barkla, Cpt. 38,49
Barkla, Isaac 14
Barnett, William J. 183
Barossa Valley 17,38,134

Barrett Brothers 205
Barrier Ranges 136-138,145
Bartle, Phillipa 195
Bastian, Henry 34,41,178
Bastian, Jack 126
Bath, Cpt. 35
Bathurst 111,112,119
Batten, Edward 40
Batten, William 13
Bayley, Arthur 176
Beaglehole, William Henry 106
Bedford Foundry 45,46
Bedruthan 208
Begelhole, William 183
Begolhall, William 114
Bell's Reef 128
Bendigo 113,114,116-118,122,124,129,
   131,132,137,144,176,202
Bendigo Miners' Association 130
Bennett, Cpt. 145
Bennett, John 170
Bennett, Nicholas 36
Bennett, Thomas 120
Bennetta, William 84,85
Bennetta, William (2) 172
Bennetts, James 13,85,167
Berryman, Jane 199
Berryman, William 186
Besanco, Henry 36
Bet Bet Creek 126
Bible Christian Missionary Society 80
Bice, John George 107
Biscovey 14,18,196
Bishop, John 167
Blackett, John 12
Blatchford, James 80
Bligh, William 9
Blight, William 41
Blinman 50
Bloemfontein 198
Boaden, Nicholas 20
Boase, Charles 119
Boksburg 198
Bodmin 12,13,20,35,52,74,111,119,172
Bokiddick 42
Bolitho's Smelting Works 37
Bone, James 41,113
Bone, Jane 41
Bone, William 41
Bonython, John Langdon 69,70,193,
   197,200,211
Boobyalla River 191
Booleroo 178
Boscarne 20
Boscastle 147
Boscaswell 131
Bosigran 120

Boswarva, Mr. 36,95
Botallack 20
Botany Bay 8
Boucaut, James Penn 28,69,70,92,93,98, 99,104,105,107,200
Boulder 177,180-182,184,185
Bowden 178
Bowden, Charles Wesley 44
Brandy Creek 191
Bray, Emma 139
Bray, J. 101
Bray, John H. "Dancing" 76
Bray, Joseph 43
Bray, Samuel 21,41,43,52,79
Bray, William 43
Bray, William (2) 139
Bray, William (3) 196
Breage 8,20,36,74,189
Bremer River 38
Brisbane 191
Broad, Mary 9
Broken Hill 77,88,94,102,104,135,140-148,165-172,174,177,181,184,185,188, 194,196,197,202,209
Brokenshire, John 181
Brookman, W.G. 176
Brunel, Isambard Kingdom 120
Bryant, Matthew 35,39,113,114,132
Bryant, William 9
Bryant (O'Bryan), William 140
Bryant, William Treffry 50
Budock 52
Bulldog Gully 126
Bulldog Creek 114
Buninyong 113
Burnard, James 147
Burr, Thomas 94
Burra Burra 16,24,35-37,40-44,51,70-73, 76-83,85,89,92,95,103,105,112-115,117 119,120,124,127,128,131,134-137,140-142,147,148,165,166,169,178,183,188, 191,192,196
Burrows, Jacob 165,167
Burtt, Thomas 77
Butte City 52

Calenick 37
California 90,111,116,117,124,128,145, 176,195
Callington 52,107,116
Callington (S.A.) 38,43,72,112,114,132
Calstock 14,45,167,184,200,201
Camborne 14,20,24,29,33,73,84,90,119, 124,126,129,138,139,143,145,167,200, 206
Camelford 14,127
Campbell's Creek 112
Canada 197
Canadian Gully 117

Cape Yorke 193
Caradon Hill 185
Carlyon, Angus 17,119
Carlyon, William Pomeroy 119,120,121
Carthew, John 142
Carthew, Stephen 33,113,143
Castlemaine 113,119,126
Chacewater 13,34,39,87,113,120
Chapell, Alfred 143
Chapman, John 41
Chapman, William 41,170
Chappel, William 44
Charleston, David Morley 106,107
Charters Towers 187,192
Chellew, Arthur 120,121
Chenoweth, William 127
Chenery, Alfred 126
Child, Parson 189
Chile 197
China 193
Chilsworthy 190
Christopher, William 34,114
Clare 143
Clunes 112,127,130
Coad, James 198
Cobar 194
Cobbler's Gully 113
Cock, F. 82
Cock, Thomas 95
Cocking, John 180
Cocking, Thomas 95
Cockburn 135,141
Coles, Mr. 52
Colwell, Charles 71
Columbia 35,197
Common, Thomas 196
Congdon, John 35,37
Constantine 36,73
Constantine, Edward 201
Conybeare, C.A.V. 90
Cooke, Ann 116
Cookstown 192
Coolgardie 175-182,185,187,188
Coon, Thomas 13
Copperhouse (S.A.) 36,71
Corbett, "Gentleman Jim" 184
Cornelius, Thomas 43,114
Cornish, John 39
Cornish Association of Broken Hill 165
Cornish Association of South Australia 69,77,105,187,210,211
Cornish Gorseth 211
Cornish Settlement 111,112,119
Cornish Town 120,124
Cornwall (Tas.) 9,190
Cornwall Central Relief Committee 8
Cornwall and Devon Society 22,69
Corsica 181

Cowling, Richard 43,44
Cowling, Thomas 42,200
Creed 14
Creswick 120,129,131
Crewes, Ernest William 147,148
Cross Roads 84,172
Crougey, Henry 197
Crowan 19,43,195
Cuba 197
Cumbria 39
Curgenwin, Samuel 169
Curnow, James 22
Curnow, John 38
Curnow, John (2) 41
Curnow, Stephen 120,128
Curtis, W. 50
Cury 199
Cwm Avon 73

Daddow, R.J. 170
Daisy Hill 114,121
Dalley, John 118
Daniel, William 120
Dare, Robert 22
Darwin 143
Datson, James 49,133,195
Davey, Charles 182
Davey, John 18
Davitt, Michael 71,104
Davy, Edwin 37
Davy, Humphry 37,47
Dead Finish 177
Deason, John 126
Deeble, Malachi 87
Devon and Cornwall Miners' Gold
    Company 118
Devon and Cornwall Syndicate 145
Devonshire 7,17,23,24,36,39,46,69,139,
    140,143,193
Diamond Rock 178
Dinkey Gully 115
Disraeli Co. 192
Doble, Cpt. 50
Doney, John 76
Downs, Henry 172
Draper, Daniel 80-82
Drew, Samuel 148
Duckham, A.B. 12,13,189,195
Duke of Cornwall's Light Infantry 198
Dunn, Charles 15
Dunn, Mary 15
Dunnolly 126
Dunstan, Cpt. 183
Dunstan, Cpt. (2) 200,201
Dunstan, Edward 45,47,73,145,194
Dunstan, Edward (2) 37,191
Dunstan, John 37,113
Dunstan, John (2) 114

Dunstan, John (3) 166
Dunstone, John 41
Dunstone, Mary 41
Dunstone, Robert 115
Dutton, Francis 25,33,34,197

Eaglehawk 118
East, Cpt. 44,45,47,76,182,184
East, J.J. 184
East Moonta 76,84,177,180
Echunga 134
Eddy, David 120
Eddy, Ellen 107
Eddy, James 145
Eddy, Philip 172
Eddy, Richard 120
Eddy, Richard (2) 120
Edwards, Thomas 207
Eliot 8
Ellery, Richard 41
Ellis, Cpt. 145
Ellis, John 119
Ellis, Richard 177
Enfield 196
English and Australian Copper Co. 37,
    106
Eucla 178
Eureka Stockade 117
Euriowie 145-147,176
Eyre Peninsula 49

Falmouth 12,13,15,21,41,52,79,118
Falmouth (Tas.) 9
Families.
    Cornish Emigrant see Appendix A 213
Faull, Christopher 87
Ferron's Reef 126
Fiery Creek 126
Fishers Gully 191
Fitzsimmons, Bob 184
Flamank, Henry 196
Fletcher, Jane 14
Fletcher, Sukey 14
Ford, William 176
Forest Creek 113,114,121,122,124,132
Forest Range 134
Foster, Joshua 86
Fowey 9
France 19,209
Francis, W.F. 183
Friar's Creek 120,127
Frost, Billy 177
Fry, A.J. 167
Fryer's Creek 112
Fullerton 145

Galapogos Islands 119
Gallipoli 208

233

Gawler 37,69,117,141
Gawler, Governor 15,146
Geake, Mr. 12
Gear 122
Geraldton 174,186
Germoe 8,74
Giles, Henry 120,121,123,128
Gill, Reuben 86,96,98,99,101,105
Gilles, Osmond 33
Glasson, Charles 14
Glasson, John 111,119
Glasson, Martha 41
Glen Osmond 16,18,33,37,128
Gluyas, Thomas 207
Glynn, Patrick McMahon 71
Golden Point 113
Goldsithney 41
Goldsworthy, Mary 24
Goldsworthy, Richard 35
Goonvrea 14
Gowling, Mr. 28,29
Goyne, William 119
Gray, Samuel 147
Great Sandy Desert 175
Gregor, Thomas 197
Grenfell, Pascoe 13
Grenfell, Richard 111
Grey, Governor 35
Grigg, John 22
Grose, Thomas 106
Gulgong 111
Gulval 118,120,122
Gummow, William 34,113
Gundagai 111,112
Gundry, Cpt. 22
Gunnislake 7,31,44
Gwennap 13,14,21,24,27,36,39,120,146,
  166,171,191,197,205
Gympie 191,192

Hall, Emma 143
Halse, John 80
Halsetown 38
Hambly, William 183
Hamblin, John 141
Hamiltons 196
Hammer, John 119
Hancock, Ellen 139
Hancock, H. Lipson 43,47,103,142,202-
  204
Hancock, Henry 139
Hancock, Henry Richard 30,39,43,46-48,
  87,97,99-101,103,107,133,140,183,188,
  200,202
Hancock, Leigh 183
Hancock, Joseph 21
Hancock, R. 101
Hancock, Richard 128

Hannaford, George 184
Hannan, Paddy 176
Hannan's PTY Ltd. 188
Hargraves, Edward Hammond 111
Harris, Ambrose 36
Harris, Charles 43
Harris, J. 34
Harris, John 114
Harrowbarrow 116
Harry, William 148
Hartz Mountains 48
Harvey & Co. 46
Harvey, Catherine 143
Harvey, John 121
Harvey, Richard 120
Harvey, Thomas 178
Hathorne Davy 46
Hawke, Albert Redvers 188,207
Hawke, Cpt. 145
Hawke, George 111
Hawke, Nathaniel 40
Hawke, Thomas 207
Hawke's Foundry 37
Hawker 50,138
Hayes, W.H. 75,87
Hayle 13,37,46,107
Haywood, Johnson Frederick 78
Helston 9,12,18,39,52,73,86,106,113,115,
  119,120,138,183,200
Helston (S.A., nr Kapunda) 34
Helston (S.A. nr Moonta)
Henwood, George 8
Herbert, Mr. 207
Herodsfoot 24
Hicks, Cpt. 145
Hicks, Frances 52
Hicks Mill 39
Higgs, Richard 126
Higgs, Samuel 42,46-48
Hill End 111
Hingston Down 185
Hitchens, Josiah 147
Hobart 189,190
Hobson's Bay 119
Hocking, Cpt. 145
Hocking, John 143
Hodge, Mary 199
Hodge, William 69
Holman, John 15
Holman's of Camborne 147
Holman's Tregaseal Foundry 129
Home Rule 111
Hooper, Joseph 22
Hooper, Richard "Dickie" 106,107
Hooper, R.J. 148
Hooper, Robert 143
Hore, Silas 196
Horrabridge 39,46

Horton, Tom 182
Hosken, W.H. 84,86
Hoskin, John 69
Hosking, John 120
Hosking, Thomas 120
Hosking, W.N. 118
Howe, Lord 9
Hummocks, The 74
Hunkin, Leslie 207

Illogan 13,14,20,124,126,201
India 181
International Workers of the World 170
Ireland 13,181
Irish Land League 71
Isaac, David 143
Isaacs, John 43,168
Ivanhoe 176,188
Ivey, John 126
Ivey, Thomas 171

Jacka, Thomas 195
Jago, Cpt. 21
James, Harry 182
James, Henry 122
James, John 129
James, W.F. 87
Jeffery, Mary 206
Jeffery, James 124
Jeffery, Richard 126
Jeffery, William 126
Jenkin, John 52,113
Jenkin, John Grenfell 143,166
Jenkins, James 112
Jennings, G. 12
Jericho 42
Jewell, James 44
Jewell, William 126
Johannesburg 52,129,198
John Williams & Son 36
Johns, Adam 175
Jones Foundry 37
Jose, George 200,201
Jupiter Creek 134
Jury, Charity 80

Kaawaa 195
Kadina 29,41,45,73,76,77,84,143,144,148
    168,171,172,180,182,198
Kadina Miners' Association 98
Kalgoorlie 74,88,167,176-179,182,183,
    185-188,197,202,209
Kanmantoo 43,72
Kapunda 16,18,26,33,37,40,41,43,44,51,
    70,71-73,80,82,83,88,104,105,114,117,
    119,131,137,140,141,143,168,171,176,
    188,191,197,205
Kawakawa 195

Kawhiia 193
Kea 98,115
Kellaway, William 143
Kelynack 38
Kelynack, Job 119
Kemp, Joseph 143
Kemp, Thomas 126
Kent, William 24
Kenwyn 14,16
Kerby, William 144
Kernewek Lowender 211
Kestel, Ralph 34,113
Keweenaw 52,197
Kilkhampton 86,89,166
Killicoat, Isaac 40,196
Kimberley 175
King Arthur 7,211
King, Philip 8
Kinsman, John 30
Kitto, Collingwood 96
Klondike 198
Kneebone, Henry 187,207
Knowles, Mr. 96
Kooringa 36,71,78-82,128,143,148,192
Korong 117

Lander, John 96
Lane, Richard 111
Lane, William 111
Lands End 8
Lanhydrock 21
Lansell, George 131,176
Lanyon, James "Jammie" 131
Latchley 190
Lathlean, Joseph 180
Latimer, Isaac 12,13,15,19,111,189,195
Launceston 8,12,15,19,37,190,203
Launceston (Tas.) 9,190
Laurimer, Marmaduke 13,52
Lawry, John 120
Lawry, W. 189
Lean, Stephen 22,33,50,194
Leeds 46
Leifchild 18,26
Lelant 120
Lemon 8
Lennell, Leigh 208
Lethlean, Henry 20
Lewis, Lewis 119
Liddicoat, John 20,44
Liddicoat, Jos 182
Linkinhorne 13,19
Liskeard 34,118,144,191
Liverpool 118
Lizard, The 18,199
London 13,15,20,34,183-185
Long Gully 116
Lostwithiel (S.A.) 36

235

Ludgvan 36,113,120,122
Luke, Cpt. 139
Luxulyan 14,42

Macclesfield (S.A.) 114
Madden, William 44
Maddern, Jane 107
Madron 13,52,72,121
Maitland 51
Major, Thomas 127
Manchester, Mr. 122
Manna Hill 134
Manuel, Francis 43,194
Maquarie River 111
Marazion 26,43,80,120,175
Marble Hill 175,183
Martin, Alice 189
Martin, F. 127
Martin, John 19,37
Martin, H. 178
Martin, Thomas 195
Martin, Tobias 189
Martin, William 178
Martin's Foundry 37,146
Martyn, Elizabeth 196
Mathews, Cpt. 118
Mathews, P.C. 119
Matthews, Cpt. 144
Matthews, Peter 120,124,128,129
Matthews, Thomas 52
Matthews, William 143
Mawgan-in-Meneage 15,18,20,52
May, Alfred 146
May, Frederick 146
May Brothers' Foundry 37,146
Mayne, Elisha 45
McArthur, J. 93,99
McLachlan, A.J. 200
Medland, Peter 14,18
Medlyn, Thomas 73
Melbourne 115-119,133,192
Melbourne Gold General Mining
    Association 118
Menadue (S.A.) 42
Mengoose 14
Menheniot 191,198
Menzies 175,182
Menzies, Robert 130,131
Mevagissey 21,52,75
Mexico 116,118,176,197
Mildren, William 115
Mill, James 20
Millbrook 184
Mining Managers' Association of
    Australia 144
Mines, Cornish See Appendix B 215
Mines, Australian See Appendix B 215
Mines, Other See Appendix C 216

Mitchell & Osborne 129
Mitchell, John 148
Mitchell, Samuel 174,186
Mitchell, William 35
Mitchell, William (2) 76
Mitchell, William (3) 126
Molesworth 8
Molyneaux River 195
Moonta 27,29,41,43,44,46,73-75,77,78,
    84-88,92,93,106,129,130,132,133,139,
    141-144,147,166-172,175,177-179,182-
    184,186,188,190,192,194-198,200-203,
    208-210
Moonta Camp (W.A.) 181,182,184
Moonta Copper Recovery Company 209
Moonta Mines (settlement) 30,44,73,74,
    76,83,84,107,143,167,182,187,191,201,
    204,208,210
Moonta Mines Development 209
Moonta Miners' Association 30,93,98
Moonta Mining Co. 27,30,97,99-101,132,
    133
Moonta Prospecting Syndicate 209
Moonta Town (Broken Hill) 148,181
Morcom, Cpt. 34,80
Morcom, S. 199
Mortonhampstead 39
Morish, William Henry 144,169,194
Morrison, Pearson 50
Morton, James 182
Morvah 113,120
Mount Alexander 112,113,115,116,124,
    127
Mount Barker 38,187
Mount Bischoff 190,191
Mount Browne 138
Mount Gipps 140
Mount Morgan 192,193
Mount Tarrengower 124
Mount's Bay 119
Mousehole 118
Moyle, Edward 43
Moyle, John P. 40
Moyle, P.G. 22
Moyle, Richard 20
Moyle, William 124,197
Moyses, Mary 41
Moyses, Thomas 41
Murchison River 174
Murrin, J. 101
Murray River 104
Mutton, Samuel 44
Mylor 70,92

Nankervis, Henry 40
Nankivell, B. 182
National Labor Party 187,188,206,
Neales, John Bentham 193

236

Nelson, Lord 9
Nevada 147
New Caledonia 196
Newcastle (N.S.W.) 194
New Residence 104
Newton, James 175
Newgate Prison 9
Newlyn 116,118
Newlyn East 167
Newtown 84
Nicholls, Albert 31
Nicholls, Edward 50
Nicholls, James 33
Nicholls, Nicholas 198
Nicholls, Richard 119
Nicholls, Robert 34,94
Nicholls, Thomas 39
Ninnes, Martha Maria 41
Ninnes, Thomas 41,79
Ninnes, Thomas (2) 172
Noble, John 43
Norfolk Island 9,10,15
Norseman 182
Northampton 174
North Hill 41
Northey, Alf. 182
Northey, John 126
Norwood 184
Nullarbor Plain 178,182

Oates, John 15,18
Oates, John (2) 44
Oates, Richard 126
Oates, William 145,176,186
Odgers, Josiah 115
Odgers, William 114
Olds, John 172
Oliver, Daniel 127
Oliver, William 40
Oliver, William (2) 143
Oodnadatta 177
Onkaparinga River 110
Ophir 111,112
Orange Free State 198
Orchard, Joseph 15,20,52
Orchard, Philip 191
Osborne, Cpt. 96,196
Osborne, Samuel 35
Otago 195,196
Oven River 117,129

Padstow 7,34,113
Pake-A-Range 195
Papua New Guinea 196
Palmer River 192-194
Parker, Mr. 126
Parkin, Cpt. 49
Parkyn, Thomas 181

Pascoe, Cpt. 38,50
Pascoe, Jack 182
Pascoe, J.B. 33
Pascoe, Peter 119
Pascoe, Thomas 183
Pascoe, William 42
Pate, Janet 143
Paull, Cpt. 38,49,50
Paull, John 14
Paull, Henry 51
Pawley, W. 126
Paynter, George Henry 166
Paynter, Sophia 39,80
Paynter, Thomas 36
Paynter, William Arundell 39,49,80,166
Peak Downs 191,194
Pearce, George 81,92
Pearce, George (2) 187,188
Pearce, Harry 184
Pearce, Sam 176
Pearce, Thomas Gilbert 175,183
Pearce, William 40
Pedler, James 13
Pedler, John 206
Pedler, William 115
Pellew, John 80
Pellew, Thomas 80
Pellew & Moore 148
Penberthy, Cpt. 38,50,145
Pendarves 8
Pengelly, J.W. 148
Pengelly, W. 71
Pengelly, W. (2) 101
Penglaze, S. 35
Penhall Roberts & Co.
Penhall, William 38
Penna, A. 95
Pennyweight Hill 113
Penrice 38
Penrice (S.A.) 38,80
Penrose, T. 197
Penrose, William Thomas 87
Penryn 19,43,52,115,116,121
Pensilva 200
Penzance 18,19,34,47,52,80,88,92,116,
    118,166,208,210
Penzance Poor Law Union 18
Pergam, Catherine 143
Perranarworthal 13,14
Perran Foundry 36,39
Perranporth 52,146,208
Perranwell 20,34,40,52,148,196
Perranzabuloe 8,14,113,52,146
Perth 174,176,178,179,185,188
Peterborough 135
Peters, Jimmy 168
Peters, Robert 115
Philips, Thomas 114

237

Phillips, Cpt. 139
Phillips, Alfred 37,49
Phillips, Joe 182
Phillips, Walter 42
Phillips, William 36
Phillps, John 113
Pilbara 175
Pinch, Henry 36,41,52
Piper, Richard 30,142,144,146,165,169
Plymouth 12,16,17,18,19,20,21,23,30,
   111,119,132,189
Plympton (S.A.) 145
Polgooth 129
Polkinhorn, Richard 127
Polkinghorne, E. 170
Polkinghorne, W. 126
Polkingthorne, Cpt. 145
Pollard, Cpt. 186
Pollard, H.C.P. 126
Pollard, Lloyd 208,209
Political Association 92,105
Polladras Downs 36
Polmear, Tom 120
Polwin 199
Ponsanooth Mutual Improvement
   Society 198
Polpenwith 73
Pool 41
Poole, James 140
Poole, Jane 139
Poole, Thomas 139
Pope, Henry 86
Port Adelaide 17,18,20-23,37,42,52,106,
   115,117,119,167,176,178
Port Augusta 107
Port Jackson 9
Port Phillip 116,118
Port Phillip and Colonial Gold Mining
   Company 130
Port Phillip Gold Mining Association 118
Port Pirie 127,141,146,209
Portreath 34,113
Portreath (S.A.) 42
Preston, Mr. 200
Prideaux, John 172
Prior, Samuel 169
Prisk, Cpt. 49,97
Prisk, John 30,98,100,105
Probus 24
Prospect 178
Prout, George 33,124,138
Prout, William 126
Prowse, William 19,52
Pryor, James 43,210
Pryor, Oswald 43,47,77,97,210
Pyap 104
Pyatt, Charles "Camborne Charlie" 142

Quick, John 117
Quintrall, John 36

Rame 43
Rand, The 52,176,198
Ranter's Gully 113
Rasp, Charles "German Charlie" 140
Read, George 120
Redruth 13,14,20,24,33-35,52,73,84,113,
   115,118,143,167,191,198
Redruth (S.A.) 36,71,76
Redruth (Tas.) 9
Redruth Total Abstenance Society 82,83
Reef Pleasant Creek 120
Repper, James 20
Renmark 167
Retallick, Arthur 198
Retallick, Cpt. 147
Retallick, James 144
Retallick, James (2) 198
Retallick, John 43,191
Rhodesia 198
Rhondda, The 129
Richard, G.A. 193
Richards, Cpt. 35
Richards, George 19
Richards, James "Fiddler Jim" 75
Richards, John 105,106
Richards, John (2) 180
Richards, Richard 206
Richards, Robert Stanley 108,206,207,209
Richards, Samuel 101
Richards, W. 118
Richards, William 24
Rishton 192
Roach, Henry 35,40,197
Roach, John 120
Roach, John (2) 126
Roach, Paul 47
Roach, Thomas 120
Roach, William 120
Roachock, P. 184
Roberts, Alex 183
Roberts, Alice 139
Roberts, Catherine 143
Roberts, Bill 182
Roberts, George 114
Roberts, Mary Jane 143
Roberts, Thomas 14,34,35
Roberts, William 139
Robins, Samuel 19
Roche 181
Rodda, Frederick 183
Rodda, M. 182
Rodda, Richard 17,38,41,80,119
Rodda, Thomas 49
Rogers, George 144
Rogers, J. 114

238

Rogers, S. 50
Roscrow, Richard 36
Rose, Benjamin 44
Rosewall, Walter 178
Rounsevell, William Benjamin 105
Rowe, Charles Thomas 180
Rowe, G.E. 185
Rowe, J. 101
Rowe, James 71,82,88
Rowe, John 21
Rowe, John (2) 37,38,105
Rowe, John (3) 40
Rowe, John (4) 143
Rowe, T. 145
Rowe, W. 182
Rowe, William 21
Rowe, William (2) 168
Rowett 114,143
Rowland, Tristram 20
Royal Geological Society of Cornwall 16,
    25,42
Royal Institution of Cornwall 211
Ruan High Lanes 13
Rule, Joe 112
Rule, Charles 40
Rundle, James 17
Rundle, Joseph 180

Sailor's Gully 112,126
Saltash 35,52,81,116,200
Salt Lake City 197
Sampson, James 41
Sampson, John 130
Samson, Henry 20
Samson, S. 101
Samuels, Jim 182
Sancreed 120
Sanders, Robert 28,40,83
Sandhurst 117,131
Sando, Anthony 44
Sandow, William 113
San Francisco 106,117,119
Santo, Philip 35,52,81,92,104,107,116
Sara, George 148
Sara & Dunstan 148
Saunders, Clara 185
Saunders, Philip 175
Sawle, James 16,52
Sawpit Gully 122
Scaddan, Jack 187,188
Scoble, Uriah 16
Scorrier 143
Scotland 89
Scown, Thomas 15
Scown, William 143
Sebastopol 77
Semmens, John 122
Ships, see appendix C 216

Siberia 177
Sicily 138
Silverton 138-143,145,147
Simmons, Charles 191,198
Simmons, Mary Ann 121
Sithney 120
Skewes, R. 50
Skewis, James 117,124
Slee, John 172
Sleep, Robert 127
Sllep, Thomas 15
Sleep, Samuel 141,144
Sloggett, Richard 197
Smith, Tom 182
Snell, George 143
Snell, J. 50
Snell, Richard 117
Snowy River 131
Somerville, Phyllis 210
Southampton 23
South Africa 8,30,176,183,198
South Australian Company 12
South Australian Mining Association
    (SAMA) 17,22,24,28,35-37,39,40,49,
    78,80,81,94, 95,103,132
South Australian Prospecting Association
    138
South Petherwin 15
Southern Cross 176,186
Sowden, William J. 86
Spargo, Peter 16
Spargo, William 17
Spence, W.G. 102
Spring Farm 41
Spring Gully 126
Spry, James 37
St. Agnes 14,15,36,42,43,105,126,208
St. Ann's Chapel 200
St. Aubyn 8
St. Austell 12,13,17,34,38,39,43,49,80,
    87,88,114,118,119,128,143,165,189,
    191,196
St. Blazey 13,38,43,46,112,119,143
St. Buryan 88
St. Cleer 8,17,143
St. Columb Major 7
St. Day 46,120
St. Dennis 189
St. Dominick 187,192
St. Erth 86,106,140
St. Hilary 120
St. Ives 118-120
St. Ives (S.A.) 38
St. Just-in-Penwith 7,17,28,29,41,106,
    111,118,121,126,129,143-145,148,175,
    176,185,186,196
St. Keverne 200
St. Levan 13

St. Mabyn 36,52
St. Merryn 49
St. Michael's Mount 175
St. Patrick 185
St. Petroc 7
St. Piran 7,72,210
St. Teath 50
St. Tudy 9
Stannary (Tinners') Parliament 7
Stannary System 32
Stannifer 190
Stannum 190
Stephens, H. 101
Stephens, John 9,12,92
Stephens, S. 101
Stephens, Samuel 12
Stenalees 196
Stevens, Cpt. 145
Stithians 19,37,113,120,125,143
Stocker, Willam 196
Stockyard Creek 126
Stoke Climsland 14,16,34,80,114
Stokes, Ralph 129,169
Strathalbyn 38,39,46,80,174
Sulky Gully 113
Swansea 34,36
Sydney 9,111,112,118,145,170,192
Sydney Cornish Association 211

Taibach 73
Tamar River 7,8,74,184
Tamar River (Tas.) 9,190
Tamar Valley 14,25,46
Tamar Valley (Tas.) 191
Tambling, Nicholas 114
Tamblyn, John 43
Tarcoola 134
Tarrengower 128
Tavistock 45
Taylor, John 39
Teague, Amelia 41
Teague, Thomas 114
Teddy, Luke 81
Terowie 141,147
Terrell, Cpt. 50
Texas 116
Thackeringa 137,138,142,144
Thomas, Cpt. 145
Thomas, Charles 146,147
Thomas, George 120
Thomas, Gilbert 200
Thomas, James 36,113
Thomas, John 126
Thomas, Josiah 168,171
Thomas, Mathew 120
Thomas, Richard "Dick" 144,165
Thomas, W.R. 165,168
Thorne, John 29,30,74,87,139,140

Timor 9
Tom, William (Snr.) 111
Tom, William 110,111
Tonkin, Cpt. 49,112
Tonkin, Absolom 38
Tonkin, John 193
Tonkin, Richard 145
Torpoint 13,184
Towednack 13,41,79,117
Toy, Edward 119
Tozer, Susan 143
Transvaal 198
Trebilcock, Merts. 182
Tredinnick, Cpt. 191
Treen 120
Tregea, John B.
Tregoning, Methusaleh 43
Tregony 9
Tregoweth, T. 51,145
Tregrehan 17,119
Tregurtha, James 177
Trelease, William 44
Treleaven, Thomas 20
Treloar, C.R. 178
Treloar, Francis 20,115,122,123
Treloar, John 113,138,183,184
Tremaine, Bill 131
Tremar 17
Trembath, Arthur 172,184
Trembath, James Warren 144
Trembath, William 40
Tremenheere, Seymour 16,25
Tremewan, James 36
Tremough 116
Trenear, Henry 193
Trenery, Philip 40
Trenerry, George 167
Trenomon, Sydney 147
Trenowith, John 17
Trenwith, Art. 208
Treseder, Stephen 122
Tresise, Charles 86,139,140
Tresize, Cpt. 145
Tresize, Alfred 20
Tresmeer 13
Trestrail, Cpt. 97
Trestrail, James 39
Trestrail, Thomas 143,185
Treveal 120
Trevean, Joe 95
Trevellyan, John 125
Trevena, William 34,114
Trevessa 117
Trevorrow, William 36
Trewartha, James 95
Trewellard 126
Trewen 15
Trewenack, Elizabeth Jane 107

Trewidden, Edward 29
Trewidden, Grace 29
Trewin, Henry 143
Trewin, James 166
Trezona, James 166
Triplett, Elizabeth 143
Trollope, Anthony 45,73,84
Troon 118
Truran, Samuel 42
Truro 12,13,16,29,34,87,112,114,118,
  119,122,144,148,165,194,195
Truro (S.A.) 38
Truscott, Charles 182
Truscott, Henry James 34
Trystan 7
Tuapeka 195
Tuckingmill 37,206
Tuckingmill (S.A.) 42
Tungkillo 37,72,107,114
Turnpikegate 18
Turon 111,112
Tyrell, Cpt. 34
Tywardreath 13,24,43

Ulooloo 134
United Labor Party 104-107,170,205-207
United States of America 15,25,26,29,50,
  118,197,198
Upper Goulburn River 126
Uren, Eliza 167
Uren, Jonathen 195

Vaundry, Cpt. 183
Verco, James Crabb 52,92,116,197
Vercoe, George 43,49,197
Verran, John "Honest John" 108,205-
  207
Verran, Stan 182,207
Veryan 20
Visick, John 98,105
Vivian 8
Vivian, G.W. 113
Vivian, A. Pendarves 90
Vivian, John 80
Vosper, F.C.B. 187,192

Waipori 196
Wakatipo 195
Wakefield, S.R. 133
Wakefield System 9
Wales 89,133,139
Wallaroo 27,37,41,43,44,46,49,73,75,84,
  93,106,132,133,141,144,168,171,178,
  182,186,188,190,194,196,197,201,207,
  208
Wallaroo and Moonta Mining and
  Smelting Co. 184,209
Wallaroo Mines (settlement) 30,73,83-86,

130,148,167,172,187,191
Wallaroo Mining Co. 27,30,99,132,133
Walter, Richard 42
Warmington, Eneder 47,96,97
Warmington, James 96,97
Warmington, William 96,97
Warn, Thomas 191
Warren, John 48,144,146
Warren, Richard 148
Warren, Tom 183
Warwick, Jack 182
Waterhouse, George Marsden 92
Waters, Henry 34
Waters, Henry (2) 114
Waters & Trevaskis 40
Wattle Flat 122
Waukaringa 134
Way, James 80
Wearne, William 121
Welcome Nugget 126
Welcome Stranger 126
Wellington 195
Wendron 38,43,52,146,210
Wesley, John 7
Wesley, W.H. 190
Wesleyan Mutual Improvement Society 82
Whitburn, Elizabeth 143
Whitburn, James 174
White Hill 116
White, John "Cranky Jan" 126
White, Mathew 120
White, William 145,190
White Rock 73
Whitford, James 114
Whitford, John 114
Whitford, Stanley R. 182,187,207
Whitford, William 114
Wilcocks, J.B. 12,17,21-24,27,132
Will, J. 69
William West and Co. 46
Williams, Cpt. 21
Williams, Cpt. (2) 139
Williams, Arthur 185
Williams, Daniel 178,179,182
Williams, David 178,179,182
Williams, John 120
Williams, Joseph 120
Williams, Philip 120
Williams, T. 101
Williams, Thomas 39
Williams, W. 145
Williams, William 120
Willowie 178
Willunga 126,127
Wilton, W. 46
Winter's Flat 113
Wisconsin 197
Withington, S. Trethewie 86

241

Wombat Hill 124
Wooley and Nephew 133
Worthing Mining Co. 37
Wright, John G. 82,86

Yass 111
Yelland, Thomas 36
Yelta 43,84,172
Yeoman, R. Carlyon 20,87
Yilgarn 186
Yorke Peninsula 20,26-28,30,31,39,40,44-
46,49,51,70,73,74,85,87,89,103,105,
131-140,145,168,169,171,172,177-179,
181,182,184,187,191,195,202,205,207,
208,210,211
Yorke's Peninsula Gold Mining and
Prospecting Co. Ltd. 182
Young, Governor 22
Yudnamutana Copper Co. 50

Zeehan 191
Zennor 120